Lignocellulosic Materials

Lignocellulosic Materials

Managing Editor: A. Fiechter

with 49 Figures and 18 Tables

Akademie-Verlag Berlin 1989

Die Originalausgabe erscheint im Springer-Verlag Berlin Heidelberg New York
als Volume 38 der Schriftenreihe
Advances in Biochemical Engineering/Biotechnology

Vertrieb ausschließlich für die DDR und die sozialistischen Länder Akademie-Verlag Berlin

ISBN 3-540-50163-0 Springer-Verlag Berlin Heidelberg New York Tokyo
ISBN 0-387-50163-0 Springer-Verlag New York Heidelberg Berlin Tokyo

ISBN 3-05-500662-3

Erschienen im Akademie-Verlag Berlin, Leipziger Straße 3—4, Berlin, DDR-1086
Lizenznummer: 202 · 100/534/89
Printed in the German Democratic Republic
Gesamtherstellung: VEB Druckerei „Thomas Müntzer", 5820 Bad Langensalza
LSV 1315
Bestellnummer: 763 973 3 (3070/38)
12800

Table of Contents

Production of Ethanol from Lignocellulosic Materials Using Thermophilic Bacteria: Critical Evaluation of Potential and Review
L. R. Lynd 1

Advances in Lignocellulosics Hydrolysis and in the Utilization of the Hydrolyzates
F. Parisi 53

Modelling, Identification and Control of the Activated Sludge Process
St. Marsili-Libelli 89

Production of Ethanol from Lignocellulosic Materials Using Thermophilic Bacteria: Critical Evaluation of Potential and Review

Lee Rybeck Lynd
Thayer School of Engineering, Dartmouth College Hanover NH 03755, USA

1 Introduction 2
2 Overview of Ethanol Production from Biomass 3
2.1 Resource Aspects 3
2.1.1 Petroleum Supply and Demand 3
2.1.2 Substrate Availability, Composition, and Potential Ethanol Yield 5
2.1.3 Environmental Impact 11
2.2 Technological Aspects 12
2.2.1 Current Ethanol Production and Utilization 12
2.2.2 Production Technology 13
2.2.3 Properties of Ethanol as a Fuel 17
2.2.4 Energetic Considerations 18
2.2.5 End-Use 19
3 Potential of Thermophilic Bacteria for Ethanol Production 21
3.1 Identification of Distinguishing Features 21
3.2 Evaluation of Distinguishing Features 25
3.2.1 General Impact 26
3.2.2 Basis for Economic Analysis 28
3.2.3 Economic Impact 30
3.3 Comparison with Other Ethanol Production Processes 34
4 Progress Toward Realization of the Potential of Thermophilic Bacteria for Ethanol Production 36
4.1 Cellulase Production and Activity 36
4.2 Utilization of Pentose Sugars 38
4.3 Ethanol Tolerance 39
4.4 End-Product Metabolism and Ethanol Yields 41
5 Concluding Remarks 44
6 Acknowledgements 46
7 References 47

Resource and technological aspects of ethanol production are considered. Conversion of lignocellulosic substrates to ethanol via thermophilic bacteria is then addressed, with particular emphasis on evaluation from an engineering perspective.

The biological conversion of lignocellulosic materials to ethanol is a versatile process which can be used in various applications for replacing or improving petroleum products, treating wastes, or reducing air pollution. Petroleum replacement can be in relation to neat fuels, fuel additives, or raw materials. Waste treatment applications may be either for wastes which require treatment (e.g. municipal solid waste) or wastes which do not (e.g. many forestry and agricultural residues). Biological treatment of solid wastes with concomitant ethanol production may become attractive in that solid wastes represent less expensive substrates than those usually considered for ethanol production. In addition, the potential energetic yield of ethanol production is about twice that of electricity generation, and electricity and ethanol have comparable value per unit energy.

Estimated recoverable oil reserves represent a resource approximately 75 times the current annual consumption on a world-wide basis. However, some countries are in a particularly poor position with regard to petroleum supply and demand. For example the U.S. estimated recoverable oil reserves represent approximately 15 times the current annual consumption. The annual ethanol production potential in the U.S. achievable within 20 years is estimated at 1.3×10^{13} MJ based on a compilation of estimates for the rates of production and availability of various biomass materials. Relative contributions to this potential are: 41 % for wastes, 39 % for energy-devoted forestry, and 19 % for energy-devoted agriculture. Notably only 6 % of the total ethanol production potential is derived from corn. Pentose sugars represent 28 % of the total potential with hexose sugars the remainder. Ethanol can displace gasoline at a ratio of about 1:1.3 on an energetic basis, thus 1.3×10^{13} MJ of ethanol can displace about 1.7×10^{13} MJ of gasoline. The U.S. ethanol production potential of 1.3×10^{13} MJ, or 1.7×10^{13} MJ of displaced gasoline, can be compared to the yearly U.S. consumption of 7.5×10^{13} MJ for energy of all kinds, 2×10^{13} MJ for the transportation sector, and 1.2×10^{13} MJ for gasoline.

Four distinguishing features of thermophilic bacteria for ethanol production in comparison to yeast systems are identified. These include the advantages of pentose utilization and in situ cellulase production and cellulose utilization, and the disadvantages of low ethanol tolerance and low ethanol yield. Many frequently-cited advantages are not considered to be of great significance from an economic viewpoint, including facilitated product recovery and high conversion rates. The economic impacts of the distinguishing features of thermophiles for ethanol production are evaluated relative to a base-case process for ethanol production consisting of pretreated hardwood hydrolysis using *Trichoderma reesei* cellulase followed by conversion of soluble hexose sugars by yeast and reaction of xylose to furfural. Relative to the base case, the impact of in situ cellulase production and substrate hydrolysis is to lower the ethanol selling cost by 37%, and the impact of pentose utilization is to lower the cost by 23 %. These two features together increase the ethanol yield per unit wood substrate by 47 % over the base case. The increased cost of ethanol separation at low concentrations appears to be relatively small if energy-efficient processes are used, however such processes have not yet been implemented on a large scale. High ethanol yields must be obtained if thermophilic ethanol production is to be practiced on a significant scale.

Research results pertaining to the distinguishing features of thermophiles for ethanol production are reviewed. Critical research areas are proposed for closing the large gap between the potential of thermophilic bacteria for ethanol production and that which has been experimentally realized to date. These include process-oriented studies utilizing potentially realistic substrates and conditions, and both biological and engineering approaches to increasing ethanol yields.

1 Introduction

Widespread recognition of the finite nature of the world's petroleum resources has led to examination of alternative sources of materials and energy. One such alternative is biological ethanol production from renewably-produced products of photosynthesis. The most abundant products of photosynthesis, also considered to be the most abundant renewable natural resource available to humandkind, are lignocellulosic materials [1]. The composition and structure of lignocellulosic materials have been reviewed [1], and the composition, availability and ethanol production potential are considered in Sect. 2.1.2.

Anaerobic bacteria with optimal growth temperatures of ≥ 60 °C have frequently been proposed for the production of ethanol from lignocellulose. Over the last ten years, rather intense research programs have been directed toward ethanol production using thermophilic bacteria in perhaps a dozen laboratories world-wide. Recent and comprehensive reviews are available which consider the cellulase enzyme complex of *Clostridium thermocellum* [2], the genetics and biochemistry of *Clostridium* including thermophilic species [3], aspects of the general and applied physiology of thermophilic bacteria [4,5,6,7,8,9,10], thermophilic cellulolytic bacteria [11,12], and thermophilic conversion of cellulose to ethanol [13,14]. Many of these reviews contain exhaustive lists of thermophiles and their properties, as well as detailed discussions of biochemistry and physiology relating to thermophiles and thermophilic ethanol production.

Typically a thorough consideration of biologically-related topics is followed by more cursory reference to engineering-related issues.

There is no attempt made here to duplicate the many excellent biologically-oriented reviews of thermophilic ethanol production. Instead this paper considers engineering topics first, and in detail, followed by a focused review of biologically-related topics. Engineering topics addressed include an overview of ethanol production from biomass, Sect. 2, which focuses on lignocellulosic materials and includes both resource and technological aspects. The potential of thermophilic bacteria for ethanol production in comparison to yeast-based processes is then considered in detail from an engineering and economic perspective in Sect. 3. Section 4 considers progress toward realization of the potential of thermophilic bacteria for ethanol production, emphasizing research areas judged to be of particular applied significance based on considerations addressed in previous sections. Section 5 offers concluding remarks. This paper is intended to be an interim progress report on the use of thermophilic bacteria for ethanol production from lignocellulose. Both review of what has been done and exposition of what could be done are important parts of this task.

Consideration of several topics, in particular ethanol production potential, will focus on the United States. This focus is convenient because it allows more specific statements to be made about issues such as substrate availability and economics. In addition, the situation with respect to petroleum supply and demand is particularly acute in the U.S., as discussed in Sect. 2.1.1. Because of this, the potential of biomass resources to replace petroleum in the U.S. has frequently been considered, albeit with different conclusions, and abundant data are available.

The rate of oil consumption in the U.S. is approximately 27% of the world-wide rate, and the highest of any country in the world [15]. However, the quantity of ethanol which may be derived from biomass in the U.S. is representative of that in other countries on a per capita basis. Vergara and Pimentel [16] give values for the annual photosynthetic energy (MJ $\times 10^3$) fixed per capita of 250 for the U.S., 250 for Sweden, 38 for India, 465 for Brazil, and 536 for Sudan at 1976 population levels. The total solar energy fixed by photosynthesis worldwide is approximately 1×10^{15} MJ a^{-1} [17], which may be compared to the 1980 world-wide annual demand for petroleum, 1.26×10^{14} MJ, and for all forms of energy, 2.9×10^{14} MJ a^{-1} [15]. At the 1976 world population of about 4 billion people, the world-wide solar energy fixed per capita was 250×10^3 MJ per capita. In 1987 this value is 200×10^3 MJ per capita.

2 Overview of Ethanol Production from Biomass

2.1 Resource Aspects

2.1.1 Petroleum Supply and Demand

A paper written in 1987 on production of ethanol from renewable resources may seem out of place. After all, oil sells for under 20 U.S.$ per barrel, and at times has been less than 1/3 of its real price in 1980. Though the fact of finite petroleum reserves is certainly felt with less immediacy today in economic terms, the long-range picture has changed very little in resource terms.

Conversion of biomass to ethanol has received attention as a means of replacing energy and materials presently derived from oil. The long-term motivation for replacing oil is the finite nature of its supply, whereas biomass is renewably available. There is little doubt that finding and developing the world's as yet unrecovered oil reserves will be progressively more difficult and costly [18]. Thus in the shorter term before oil supplies near exhaustion, the price of oil may be expected to rise, and biomass feedstocks may become more competitive. The issues of limited oil supply and the prospect of an improvement in the relative economics of biomass feedstocks compared to oil are addressed in the paragraphs below.

In a comprehensive study published in 1982, Grathwohl [15] presents estimates for the worldwide total amount of recoverable oil at 2.4 to 3.6×10^{11} metric tons, based on 40% oil recovery during drilling. As of January 1979, ~15% of this quantity had already been recovered, and 25% represented proven reserves. In 1980 the worldwide oil consumption was 3.0×10^9 t a^{-1} [15], or about 1% of the total recoverable oil per year. Thus in 1987, approximately 23% of the total recoverable oil has already been used. Because of the uneven geographical distribution of petroleum reserves, the situation is more acute in some countries than others. For example, the U.S. had used 45% of its total estimated recoverable oil by the beginning of 1979 [15], and 64% by the beginning of 1987 [15,19]. Data cited by Grathwohl for the beginning of 1979 puts the estimated total amount of recoverable oil remaining in the U.S. at 19.7×10^9 t. Deducting oil consumption since 1979 [20] results in a value of 13.1×10^9 t for the oil remaining at the beginning of 1987. This quantity represents 15 times the annual U.S. oil consumption in 1986 of 0.87×10^9 t. The U.S. Geological Survey (USGS) recently estimated the undiscovered recoverable oil in the U.S. at 4.9×10^9 t [21], an amount equal to 5.6 times the 1986 annual consumption.

Most estimates of future oil consumption for both the U.S. and the world show a

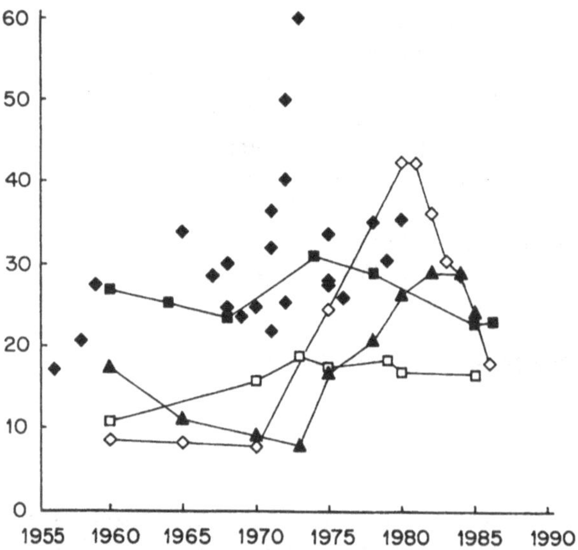

Fig. 1. Historical view of estimated worldwide oil reserves, and oil price, consumption rate, and prices of ethanol precursors for the U.S. —◆—, Estimates of recoverable oil reserves from [15] (values plotted are $10^{-13} \times$ actual values in kg a^{-1}); --◇—, oil price from [22] (1984 U.S. \$ $barrel^{-1}$); —□—, oil consumption from [15,22] ($0.5 \times 10^{-12} \times$ MJ a^{-1}); —▲—, ethylene price from [23] ($0.5 \times$ 1984 U.S. \$ kg^{-1}); —■—, price for delivered green mixed hardwood chips from [24,25] (1984 U.S. \$ t^{-1})

small increase in the rate of consumption up to the year 2000, with a decrease in consumption beginning shortly thereafter [15, 22]. The U.S. Department of Energy [22] estimates that the real price of oil in the U.S. will fall slightly by 1990, and increase by nearly 3-fold in the 20 years thereafter.

Figure 1 presents an historical view of estimates for the size of the world oil resource, and oil price and utilization rate in the United States. United States price histories are also presented for ethylene, the petroleum-derived precursor for ethanol synthesis, and delivered green hardwood chips, representative of one of the primary lignocellulosic ethanol precursors considered herein. It may be seen that estimates of the total recoverable oil reserve generally increased until 1970, but remained nearly constant between 1970 and 1980. U.S. petroleum consumption relative to 1960 had increased by 74% in 1973, and had decreased to 56% above 1960 consumption by 1980 [15, 22]. The price of oil, given in real U.S.$ [22], increased sharply in the mid 1970's and again in the late 1970's, and then fell sharply in the mid 1980's. These price changes do not reflect changes in the size of the estimated oil resource, but rather factors such as OPEC policy, and over-compensation in the development of oil pumping capacity. The real price of ethylene [23] follows the price of oil, from which it is derived, fairly closely. This is not true for the price of wood, which has a relatively stable history in the U.S. and has not increased in real terms since 1960 [24, 25]. Thus if oil prices increase as expected, it appears likely that lignocellulosic substrates such as wood will become more attractive sources of raw materials.

2.1.2 Substrate Availability, Composition, and Potential Ethanol Yield

In 1981, 4.6% of U.S. refinery output was devoted to chemical production, whereas 44.8% was used for gasoline, and 87% for fuels of all kinds [17]. In light of this difference in material requirements, several studies have concluded that large-scale utilization of biomass for chemical production, with feedstocks including but not limited to ethanol, is a realistic possibility, whereas fuel production is not realistic [26, 27, 28]. However others differ in their evaluation of the potential contribution of biomass to meet fuel needs [29, 30, 31]. For example, Ferchak and Pye [30] concluded that the theoretical limit of potential biomass resources for production of fuel ethanol in the U.S. exceeds foreseeable demand. It may be noted that the amount of ethanol required to mix with all gasoline used at 10% ethanol is comparable to the total refinery output devoted to chemical production.

Faced with this difference in evaluation of the biomass resource, it is useful to consider estimates of the size of this resource. Table 1 presents estimates for the availability of biomass in the U.S. from waste materials, forestry, and agricultural crops from 8 different studies. The three estimates for total waste are fairly close at 856 to 1081×10^6 t a^{-1}. In general, forest and agricultural residues necessary to maintain soil fertility are not included in estimates for total or collectable wastes. The two estimates for collectable waste are somewhat more disparate at 502 and 864 million tons a^{-1}. Obviously estimates for the fraction of a waste which is collectable depend on the value attached to the waste materials, which is usually unspecified. The forest production (t a^{-1}) available for energy production is variously estimated at 182 and 245 in the late 1970's, and 280 to 560 and 909 in the year 2000. The higher of the two estimates for forest production in the year 2000 is based on a greater role for high productivity energy-devoted silviculture. The availability of agricultural crops,

Table 1. Estimates for the availability of biomass in the U.S. from waste materials, forestry, and agricultural crops (millions of dry t a^{-1})[a]

	Young et al. 1986	Jeffries 1983	Ng et al. 1983	Ferchak and Pye 1981	Goldstien 1981	OTA 1980	Vergara and Pimentel 1979	Humphrey et al. 1977
Waste Materials								
Agricultural wastes + residues								
Total		350	736	382	323	381	430	364
Collectable	364		289	253[2]		75		
Forestry and mill								
Total		348	209			76		55
Collectable	91	159	155	253[b]				
Urban wastes								
Total			70	1512	146		123	136
Collectable	182		48	105[2]				
Manure								
Total			236		182		255	182
Collectable	227		10				127	
Sewage								
Total							13	14
Other							260	105
Total			1251	>909			1081	856
Total collectable	864	502						
Forestry								
Total[c]								
late 70's		379	409			616–1639		
2000 (assumes intensive management)				2727				
Available for energy production								
late 70's		182	245					
2000				909[d]		280–560		

Agricultural crops								
Corn								
1980, with by-product utilization							69	
2000, total						69–1575		
Forage Grasses								
mid 80s						91–182		
2000						0–414[f]		
Year of data	1982	1978–1981	1977–1979	1976–1979	~1976	1976–1978	1975–1978	~1975

[a] Data are from [16, 27, 30, 31, 32, 33, 34, 35)];

[b] values are for the year 2000;

[c] Estimates are based on commercial forest land available in excess of the requirements of the forest products industry; estimates also do not include wastes from the forest products industry;

[d] this value assumes intensive management and is based on large privately-owned land tracts that meet site, climate, and precipitation requirements for deciduous species silviculture and are available for energy production. This amount of land is equal to 1/3 the total forest area estimated available for energy production, and is also equal to 5.3% of the land area in the contiguous United States;

[e] the minimum value is for the case where no agricultural land is available for energy crops and production from corn with by-product utilization is at 1980 levels;

[f] full development of grass production would proclude corn production above the level where there is demand for process by-products

discussed in more detail below, is estimated at 69×10^6 t a^{-1} for corn with process by-products used for animal feed, plus either an additional 0 to 88×10^6 t of corn, or 0 to 414×10^6 t of forage grass. The studies considered in Table 1 which give values for the mid 1980's and the year 2000 were done in 1980 and 1981. The allocation of land, research, and manufacturing resources to ethanol production during the 1980's has been less than envisioned by these studies. Until this situation changes, the data in Table 1 for the year 2000 are probably more realistically regarded as biomass availability attainable in a 20 year time span, recognizing that these estimates become more approximate with time.

Ethanol production from sugar- and starch rich agricultural crops may be considered to compete with food production [36], is generally considered to have a less favorable energy balance compared to lignocellulosic substrates (see Sect. 2.2.4), and has the greatest potential for unfavorable environmental impact (Sect. 2.1.3). However, some agricultural crops are primarily used for animal feed in many countries. For example, over 80% of the total U.S. corn crop is used for feed, with 55–60% used domestically, whereas approximately 10% is used for human consumption [37]. Moreover, the starch fraction of the corn crop can be used for alcohol production while residues and/or processing by-products still retain considerable feed value [31, 37]. Thus it has been suggested that ethanol can be made from corn ultimately used for animal feed, with relatively small incremental resource demands [30, 38]. In the case where by-products from ethanol production are used for feed, corn is the most attractive of the grain and sugar crops for ethanol production [31, 37].

There are however limits to the demand for the feed materials available as by-products from corn-derived ethanol production. In a comprehensive study, the U.S. Office of Technology Assessment [31] cited values for the point where by-product utilization will drop at 8 to 11×10^9 L a^{-1} ethanol for distillers' grain, a by-product of ethanol production based on corn dry-milling [37], and as much as 26×10^9 L a^{-1} for corn gluten meal, a by-product of ethanol production based on wet-milling [37]. 26×10^9 L of ethanol correspond to about 69×10^6 t of corn abailable for ethanol production, or about 70% of the total amount of corn used for animal feed in the U.S. in the early eighties [37].

Forage grasses represent agricultural crops with a more favorable energy balance and lower potential for unfavorable environmental impact in comparison with corn [31]. Moreover grass utilization for ethanol production makes use of the entire plant, and does not depend on by-product utilization. The OTA study [31] foresees a much larger role for grasses than for corn among agricultural crops used for energy production. However, this study cautions that there is considerable uncertainty regarding the availability of agricultural land for energy crops, with between 0 and 26×10^6 ha available in the year 2000.

Table 2 presents data on the composition of representative cellulosic materials. Pentans contribute significantly to the total degradable carbohydrate in these materials. It is notable that the degradable carbohydrate content of urban waste is approximately 60% of the dry weight, and that the degradable carbohydrate content of wood, forage grass and corn are comparable when pentans are considered.

The data in Tables 1 and 2 are used to arrive at estimated values for the carbohydrate composition, annual rates of production, and potential ethanol production from biomass materials achievable within a 20-year time span, presented in Table 3. In arriving at the values in Table 3, an effort was made to choose intermediate values

Table 2. Composition of respresentative cellulosic materials[a]

	% Dry weight							
	Hardwood	Softwood	Wheat straw	Corn stalks	Forage grass	Corn	Manure	Urban Waste
Hexan								
Glucan	50	46	35	36.5		72	10–18	
Galactan	0.8	1.4	0.7	1.1				
Mannan	2.5	11.2	0.4	0.6				
Total	53.3	58.6	36.1	38.2	42	72	25	37.6
Pentan								
Xylan	17.5	5.7	19.0	17.2				
Arabanan	0.5	1.0	4.4	2.1				
Total	18.0	6.7	23.4	19.3	30			16.0
Total degradable	79.7	72.7	66.7	64.4	80.8	80.0	27.2	60.0
Lignin	21.0	29.0	14		7.0	6–10		9.5

[a] Data on hardwood, softwood, wheat straw, and corn stalks from [34]. Data on Corn from [39], data on manure from [40,41]. Data on forage grasses is estimated from [1,42,43]. Any estimate of the composition of municipal waste must be approximate. The carbohydrate compositions shown are calculated from a "typical waste" based on data from [44,45]. The typical waste contains 16.6% newspaper, 24.9% waste paper and cardboard, 16.5% yard waste, 11% food waste, and 2% wood. The cellulose hemicellulose, and lignin content of each fraction (based mainly on data from [1]) is assumed to be: waste papier and cardboard, 65/15/12; newspaper, 45/31/20; yard waste, 45/25/15; food waste, 40/25/2; wood, 55/13/24;

[b] calculation of degradable carbohydrate considers the water of hydrolysis. Thus degradable carbohydrate = (180/162)*hexan + (150/132)*pentan. Calculation of the degradable carbohydrate in manure is based on 80% carbohydrate in fiber, treating all fiber as hexan

Table 3. Estimated annual ethanol production potential for biomass materials achievable within a 20 year timespan

Material	Composition[a] % Hexan	Composition[a] % Pentan	Collectable[b] production (millions of t a^{-1})	Degradable carbohydrate (millions of t a^{-1})	Potential[c] ethanol (millions of t a^{-1})
Wastes					
Ag.	37	21	273	177	74.9
Forestry	55	17	182	146	61.8
Urban	37.6	16	91	55	23.3
Manure	20	5	136	38	16.1
Other	40	10	91	51	21.6
Total[d] wastes	38.5	15.4	773	467	198
Forestry					
	53.3	18.0	545	435	184
Agricultural					
Grasses	42	30	182	147	62.2
Corn (with by-products)	72		69	55.2	25.9
Total			1569	1104	470

[a] Values based on Table 2,
[b] values are the author's estimates based on data in Table 1,
[c] calculated based on: (degradable carbohydrate)* (0.9 degradable carbohydrate utilization)* (0.47 ethanol/utilized carbohydrate), except for corn where complete carbohydrate utilization is assumed,
[d] totals are weighted averages in the case of % hexan and % pentan

between the maximum and minimum estimates given. Biomass proponents could argue, doubtless with some validity, that these estimates are less than the ultimate practical potential. Biomass detractors could argue, also with validity, that a society requires a strong motivation to devote the resources required to produce alcohol at the levels in Table 3. The total estimated potential ethanol production from collectable wastes is 198×10^6 t a^{-1}, with the order of ethanol production potential: agriculture > forestry > urban and other > manure. The contribution of forestry other than wastes of the forest products industry is estimated at 184×10^6 t a^{-1}, representing the annual yield from 545×10^6 t of wood with the composition of hardwood. Forage grasses could add an additional 62.2 Mt of ethanol, while the ethanol potential from corn ultimately used as an animal feed is estimated at 26 Mt assuming wet-milling processing.

The total estimated annual ethanol production achievable within 20 years from renewably-produced biomass materials in the U.S. is 470×10^6 t a^{-1}, representing about 1.3×10^{13} MJ. The relative contributions of wastes, forestry, and agriculture are 42%, 39%, and 19% respectively. The contribution of pentose sugars is about 28% of the total; the contribution of corn is 6% of the total. 1.3×10^{13} MJ of ethanol may be compared to the annual usage of in the U.S. for transportation (2×10^{13} MJ) [17], gasoline (1.2×10^{13} MJ) [46] for gasoline, and energy of all kinds (7.5×10^{13} MJ) [17].

Based on the data presented in this section, it is the author's conclusion that the ethanol production potential of the biomass resource in the United States is significant relative to the current demand for liquid transportation fuel, and that this end use for ethanol cannot be dismissed based on substrate supply considerations. Thus characteristics of ethanol for use as both a fuel and chemical feedstock will be considered in subsequent sections. It may be noted that transportation fuel is the energy-related end-use for which ethanol is best suited. Most stationary applications offer little incentive for the lost available energy accompanying ethanol production from biomass substrates which could be burned directly. Future demand for transportation fuel will be affected by variables such as fuel price and availability, demographic changes, transportation fleet efficiency, and the efficiency of particular fuels [22]. The U.S. Department of Energy (DOE) has predicted that total transportation energy consumption will decline somewhat through 1995 due primarily to increased efficiency of passenger vehicles.

In a global context, the U.S. has disproportionately high oil and energy consumption and representative photosynthetic energy fixed per capita as already noted in this paper. In light of these considerations, the potentially significant contribution of biomass-derived ethanol for liquid fuel requirements in the U.S. strongly suggests that very significant potential exists in other countries as well.

The estimates in this section do not take into consideration processing losses, energy inputs for substrate growth, harvest or transportation, nor the relative fuel efficiency of ethanol and gasoline, all of which are considered in subsequent sections. Obviously these factors will affect the net work which can be derived from biomass fuels. Consideration of ethanol production potential separately from procurement, conversion, and end use related factors is useful because the latter depend to a much greater extent on the current or assumed state of several technologies than does the former.

2.1.3 Environmental Impact

The environmental impacts of ethanol production can be divided into three classes: those resulting from process waste streams; those resulting from the interplay between substrate production and land resource considerations such as soil fertility and erosion, and maintaining wildlife habitat; and those relating to ethanol utilization per se. The former class does not appear to pose problems which cannot be addressed by conventional waste treatment technology [31,47,48,49]. The principal waste streams are airborne emissions, suspended solids and BOD in wastewater, and solids in the form of ash and insoluble salts, particularly arising from neutralization of acid. Several process designs burn residual organics and sludge from wastewater treatment, thereby providing process energy and eliminating organic solid waste [50,51].

Land resource issues have sharply differing potential to be environmental problems depending on the ethanol feedstock considered. According to the U.S. Office of Technology Assessment (OTA) [31], the potential for environmental damage associated with various feedstocks is as follows. Wood and food-processing wastes, animal wastes and collected logging waste have no significant potential. Grasses should have few significant adverse impacts for most applications. Crop and logging residues have some potential for harm if mismanaged, and speculative potential for long-term damage to productivity because of loss of soil organic matter. Other wood sources have high potential but theoretically can be managed. Grain and sugar crops have the highest potential. The OTA [31] concluded that biomass has the potential to be an energy source that has few significant environmental problems and some important environmental benefits. For a number of reasons however, a vigorous expansion of bioenergy still may cause serious environmental damage because of poorly managed feedstock supplies and inadequately controlled conversion technologies. Also, some uncertainties remain about the long-term effects of intensive biomass harvests on soil fertility.

Somewhat different considerations arise for tropical regions, particularly with respect to starch and sugar crops. Matsuda and Kubota [52] have investigated the feasibility of fuel alcohol production in Southeast Asia. These authors concluded that increased land use to cultivate crops for ethanol production using field agriculture would cause large scale destruction of the ecosystem in tropical regions.

According to Grathwohl [15] it has definitely been established by measurement that the atmosphere is currently being enriched in CO_2 at the rate of 1.3 ppm per year. Increased CO_2 levels have prompted concern over climatic warming and its effects [53,54]. A very significant environmental advantage of ethanol production is that the photosynthesis-ethanol production-ethanol combustion cycle has no net CO_2 generation provided that it is driven by fuels derived from photosynthesis. A second factor relating to ethanol utilization is the substitution of ethanol for gasoline to reduce air pollution. Use of ethanol as a gasoline additive to reduce air pollution is receiving attention in urban areas of the U.S. having air quality problems related to smog formation personal communication, R. Datta, Michigan Biotechnology Institute, and [55].

2.2 Technological Aspects

2.2.1 Current Ethanol Production and Utilization

Presently, ethanol is produced either by using yeast, or by catalytic hydration of ethylene. The technology for bioethanol production is discussed in Sect. 2.2.2. Aspects of ethanol production in the U.S. have recently been considered by Venkatasubramanian and Kiem [37], Hacking [56], and Murtagh [57]. In 1977, processes using yeast accounted for 26% of the total U.S. industrial ethanol production, with the remainder made from ethylene. Between 1977 and 1982, U.S. ethanol production shifted dramatically in favor of the bioproduct [56,58]. During this period, ethylene-derived ethanol production stayed constant at $\sim 908 \times 10^6$ L a^{-1}, while bioethanol production increased from 318 to 2271×10^6 L a^{-1}. A similar trend toward bioethanol has occurred in the European Community [59]. The development of fuel ethanol production in the United States has been stimulated by substantial tax incentives [37,57]. Tax incentives are largely responsible for the difference between the recent prices of synthetic ethanol ($\sim$0.44 \$ L^{-1} anhydrous) as compared to bioethanol ($\sim$0.32 \$ L^{-1} anhydrous). Actual production costs for ethylene- and bioethanol are more nearly equal than these prices would indicate, and both are strongly influenced by the cost of substrate (see Sect. 3.3). Demand for ethylene by the chemical process industry has been near capacity levels in the U.S. in 1987 [60].

Presently synthetically-produced ethanol is used in the U.S. for synthesis of acetaldehyde and acetic acid (43%), cosmetics and pharmaceuticals (28%), cleaning preparations and solvents (16%), and coatings (13%) [56]. Ethanol for fuel use is currently produced at the level of about 3×10^9 L a^{-1} [57], and accounts for the bulk of biological production. Originally fuel ethanol was developed as a gasoline extender, but today is utilized primarily as an octane enhancer.

2.2.2 Production Technology

The technology for biological production of ethanol from biomass can be divided into 6 steps: 1) substrate growth, harvesting and transport, 2) substrate pretreatment, 3) substrate hydrolysis, 4) biological conversion, 5) product recovery, including by-products, and 6) waste treatment. Some steps can be combined or eliminated depending on the process. The emphasis here will be on ethanol production technology relevant to lignocellulose utilization, including both technology in use and in the research stage.

Substrate growth and harvesting have been considered in detail for wood by Smith and Corcoran [61] and also by Ferchak and Pye [62]. Energy requirements for growth, harvest, and transportation of lignocellulosic substrates and related equipment are discussed in Sect. 2.2.4.

Insoluble substrates for ethanol production must usually undergo some pretreatment prior to biological conversion to, at least, provide enough surface area for enzymes to obtain access to the substrate. Pretreatment processes for lignocellulosic materials are the subject of reviews by Datta [63], Wilke et al. [64]. Grethlein [65], and Dale [66]. For lignocellulosic materials, the lignin/hemicellulose/cellulose matrix has low porosity and is therefore resistant to attack by enzymatic hydrolysis [67]. Most successful pretreatments rely on removing either hemicellulose or lignin to create a material with porosity sufficient to allow significant access for enzymatic attack. Using pore-size measurements, Weimer and Weston [68] determined that a minimum pore dimension of approximately 43 Å is required for hydrolysis to be catalyzed by cellulases of both *Trichoderma reesei* and *C. thermocellum*. Cellulose crystallinity has an effect on reaction rates, but only becomes a factor once the enzyme has access to the cellulose fiber. Allowing this access is thought to be the primary function of pretreatment [69,70].

As reviewed by Dale [66] and Grethlein [65], a variety of pretreatments have been studied. Some increase in the rate and yield of enzymatic hydrolysis has been obtained using acid, base, ammonia (subcritical and supercritical), solvents, heat, explosive decompression, and combinations of these. Several pretreatments allow subsequent enzymatic hydrolysis yields of $>90\%$ theoretical. Capital and operating costs for lignocellulose pretreatments such as dilute acid hydrolysis and steam explosion are a small fraction of total ethanol production costs [51,70,71]. Thus the importance of the choice of pretreatment system is primarily the effect of the chosen technology on other process steps [65,70]. An important limitation of many pretreatment technologies is their ineffectiveness against softwood substrates [67,69,72]. However, progress is being made in this area [73].

Hydrolysis of substrates prior to biological conversion can be achieved by pretreatment followed by enzymatic hydrolysis, or by acid hydrolysis. The practicality of acid hydrolysis is hindered by glucose yields $\leq 60\%$ at low acid concentrations, and high costs for acid recovery at high acid concentrations [71,74]. Enzymatic hydrolysis achieves high substrate conversion yields, but cellulase production is very expensive. Wright et al. [70], calculate that the combined costs of cellulase production and enzymatic hydrolysis are responsible for 46% of total capital investment and contribute 0.28 \$ L^{-1} to the cost of producing ethanol from wood. Furthermore cellulase production requires substrate, about 12—15% of the total [51,70], which could otherwise be used to produce ethanol. Enzymatic hydrolysis has been reviewed by Wilke et al. [64], Ladisch et al. [74], and Mardsen and Gray [75]. Acid hydrolysis is discussed in reviews by Grethlein [76], Ladisch et al. [77], and Wright and Power [78].

Acid hydrolysis gives rise to sugar degradation products in amounts approximately equal to undegraded sugars [79,80]. Furfural, a by-product of xylose degradation, could be used as a chemical feedstock, and has received attention as a by-product to improve the economics of acid hydrolysis-based ethanol production [50,81]. The price of furfural will depend on its volume of production [81]. It has been estimated [50] that the price of furfural at the level of production resulting from 15–20 large-scale acid hydrolysis-based plants would be 22–33 cents kg^{-1}, or roughly half the price of ethanol. The yield of furfural from xylose during acid hydrolysis of biomass, ~38% on a mass basis [50], is lower than the ethanol yields which could be obtained from xylose via biological conversion. In addition there are added costs for capital, utilities, and raw materials associated with furfural production. Considering these factors, it appears that xylose conversion into ethanol is more attractive than conversion into furfural.

Enzymatic hydrolysis and biological conversion need not necessarily be separated, and combining these steps offers the potential for relieving end-product inhibition by products of enzymatic hydrolysis, and lowering capital costs. Simultaneous saccharification and biological conversion using thermophilic bacteria is the subject of Chapters 4 and 5 in this report. This approach has also been studied using ethanol-producing mesophiles in combination with added cellulase, as reviewed by Wilke et al. [64].

Both biological and engineering aspects of the biological conversion step of ethanol production are the subjects of a large body of literature (see [58,64,82,83,84,85] for reviews). Ethanol production per se is relatively inexpensive. Thus, like pretreatment, the choice of process system is based primarily on the impact on other process steps. Considering the entire process, the most important consequences of the choice of the biological conversion system are the range of substrates which can be utilized, the ethanol yield, and the ethanol concentration produced, all primarily properties of the organism and not bioreactor configuration. Much research has been devoted to increasing volumetric productivity. Approaches include techniques to keep cells in the bioreactor, principally by immobilization [86] and recycle [84]; and/or to keep the ethanol concentration low by removal via solvent extraction, either with [87,88] or without [89,90] separation of solvent and aqueous phases by a membrane, or removing an enriched vapor stream [64,85,91,92]. Batch process systems presently used for ethanol production in the U.S. typically have volumetric ethanol productivities of about 3 $g\ L^{-1}\ h^{-1}$ [37]. Continuous stir-tank reactors offer somewhat higher productivities than batch systems [84], and high productivity systems employing immobilized cells or ethanol removal can achieve productivities of 100 $g\ L^{-1}\ h^{-1}$ [37,91]. Though it has been claimed that high productivity processes offer the potential for large reductions in costs related to biological conversion for yeast-based systems [64,85], very few continuous processes are in use commercially [37,84].

Ethanol production is normally carried out by one of several species of yeast. The bacterium *Zymomonas mobilis* has also been considered (see [93,94] for reviews), as have thermophilic bacteria (see Sects. 3 and 4). Both the yeasts normally employed for ethanol production and *Z. mobilis* grow only on mono- and di-saccharides, and both can produce ethanol at concentrations up to about 10% [59]. The merits of these organisms for ethanol production have been compared [59,95]. *Z. mobilis* does not use pentoses. Of the yeasts that do use pentoses, none do so with rates and ethanol yields which are either comparable to those on hexoses or sufficient to allow a practical process [10, 70]. Efficient conversion of pentoses to ethanol is an active field of research (see [34, 42, 96] for reviews).

Recovery of ethanol from spent medium is normally accomplished by distillation, though alternative processes have been proposed, many with lower energy requirements then conventional distillation [97, 98, 99]. A variety of techniques have been proposed for dehydration [62, 100]; molecular sieve dehydration appears to be particularly promising [57, 101]. Current practice for energy-efficient distillation of ethanol is based on vapor recompression heat pumps operating between the overhead vapor and the reboiler, and also heat integration between columns operating at different pressures [102, 103]. Busche [103] reports an equivalent heat requirement (q + 3*w) of 4.24 MJ L^{-1} for producing essentially pure ethanol from a 10 wt. % feed using state of the art distillation techniques.

Extractive distillation offers the dual advantages of low reflux ratios, and therefore low energy requirements, and also elimination of the azeotrope. Ethanol recovery by extractive distillation has been studied by Barba et al. [104] and also by Schmitt and Vogelpohl [105]. Both studies concluded that this technique offered the potential of significant energy savings. Solvent extractive agents are effective [106], however their use involves very large column temperature drops and so does not favor heat integration between columns. Lynd and Grethlein [107] have designed an ethanol distillation process specifically for separating ethanol from dilute broths. This process uses intermediate heat pumps with optimal sidestream return, IHOSR [108, 109], in conjunction with extractive distillation. A flow sheet for a modified version of this process is presented in Fig. 2. Capital costs for this process and the impact on the ethanol selling price of substituting this process for conventional distillation are presented in Sect. 3.2.3. In the modified version condensation of the overhead vapor from the extractive column provides heat for the evaporator and the stripping column, which are operated at a lower pressure. The main advantage of the modified version over the original process design is that the stripping column is the low temperature column instead of the high temperature column. This property decreases or eliminates heat required for preheating the feed to the column temperature. In addition, the temperatures in the stripping column can be made low enough that cells may pass through the column, which was not practical in the previous design.

The energy requirements (q + 3*w) for the IHOSR/extractive process are shown in Fig. 3 as a function of feed ethanol concentration assuming a saturated liquid feed. Also shown in Fig. 3 are energy requirements for conventional distillation. It may be seen that the energy requirement of the IHOSR/extractive design remains relatively flat down to concentrations as low as 1 wt. % ethanol, whereas that for conventional distillation increases sharply with decreasing concentration.

General economic issues related to ethanol production have been reviewed primarily for production from corn [37, 56, 110, 111], and for production from wood [70]. Some features of ethanol production economics are of importance to subsequent sections in this paper. The cost of production of bioethanol is dominated by substrate costs and capital-related costs. In the manufacturing cost summary presented by Venkatasubramanian and Kiem [37] for ethanol production from corn, the substrate cost, including by-product credits, represents 46 % of manufacturing costs, capital-

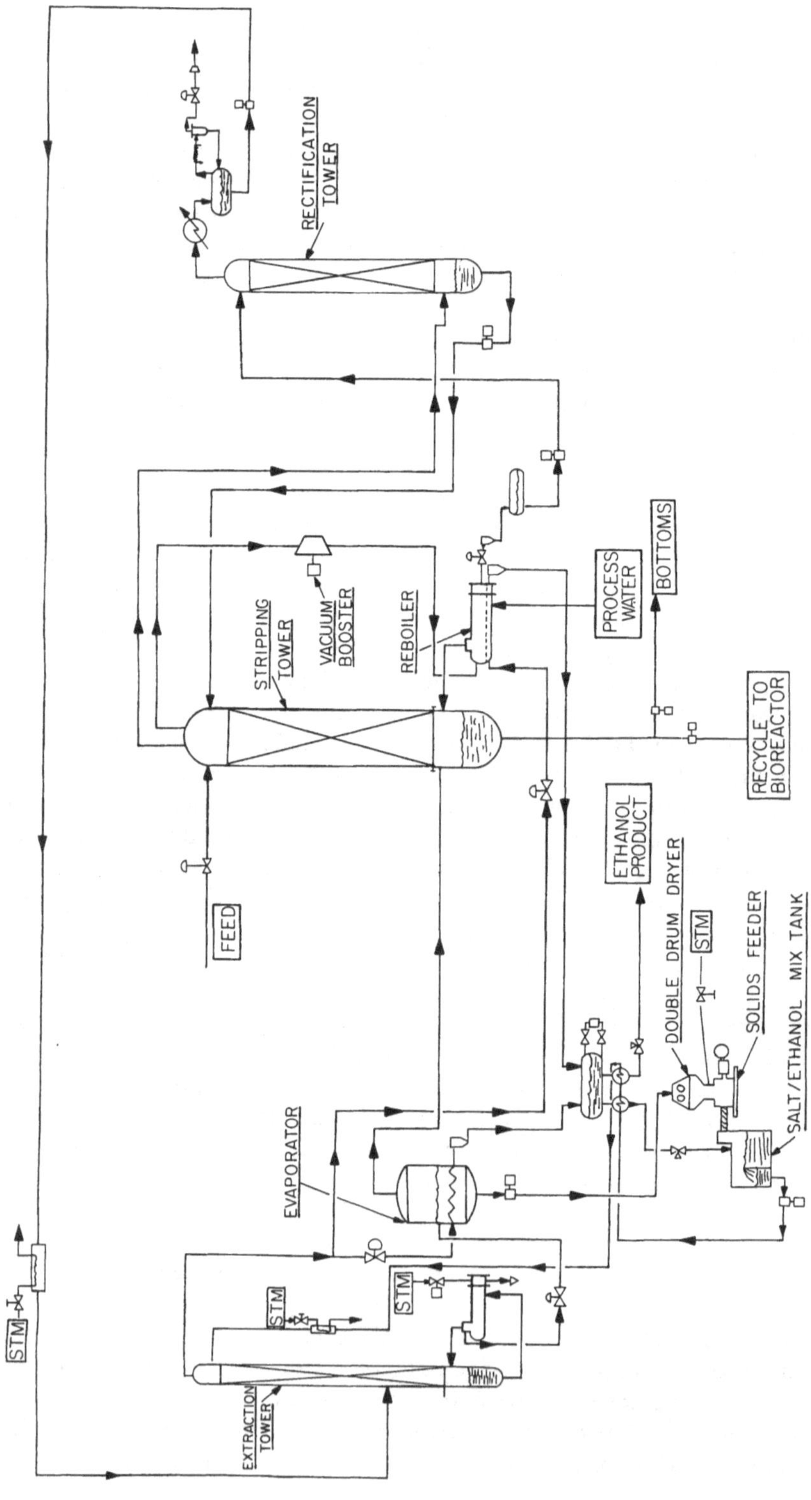

Fig. 2. Flow sheet for ethanol separation using IHOSR/extractive distillation (modified from [107])

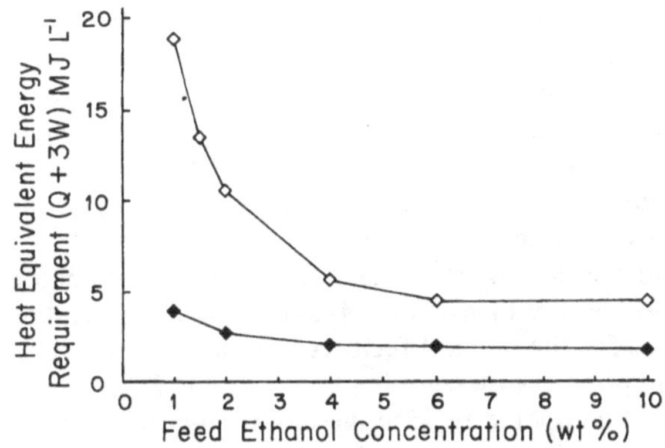

Fig. 3. Energy requirements (q + 3*w) as a function of ethanol concentration for anhydrous ethanol production using conventional distillation —◇—, and IHOSR/extractive distillation —◆—

related fixed expenses represent 33%, and variable expenses including utilities represent 21%. A detailed economic summary of ethanol production from hardwood using yeast and enzymatic hydrolysis and also an economic summary of yeast-based ethanol production from corn are presented in Sect. 3.3. Waste treatment for ethanol production processes was briefly considered in Sect. 2.1.3.

2.2.3 Properties of Ethanol as a Fuel

The debate over fuel ethanol has focused on issues of substrate supply, production economics, and net energy balance. In contrast with the widely differing schools of thought on these matters, it is generally agreed that alcohols make excellent fuels for spark-ignited engines [31, 112, 113]. A few considerations are relevant here.

The energy density of ethanol is 65% that of gasoline (based on low heating values, used throughout). However, several factors, described below mitigate this difference both for use of ethanol in gasoline mixtures and as a neat fuel. Moreover, it is not clear that energy density should be an overriding consideration when evaluating alternative fuels. The energy yield, work done or distance traveled per unit energy, would seem to be at least as important.

Ethanol has a higher octane rating, $(R + M)/2 = 102$, than does gasoline, and causes a disproportionate increase in octane when blended with gasoline. The blending $(R + M)/2$ octane rating of ethanol is 119. This may be compared to values for other octane-enhancers: 120 for methanol, 97.5 for tert-butanol, and 108 for methyl tert-butyl ether [113]. Methanol boosts octane as effectively as ethanol, but blending methanol with gasoline results in greater difficulty with phase separation, materials incompatability, and vapor lock [112, 113]. Estimates for the volumetric milage for ethanol blended 10% with gasoline vary from 96% to 107% relative to gasoline [31, 114]. An intermediate value considering the range of estimates is equivalent volumetric milage [114].

Burning ethanol as a neat fuel requires carburetor modification and provision for cold starts, as well as minor changes in engine materials. These are not difficult problems in a new system, but make retrofitting somewhat awkward [113]. Automobiles designed for burning hydrous ethanol have been marketed in Brazil by Fiat, Ford, Volkswagen, G.M. and Chrysler [62]. Engines designed for ethanol can have higher compression ratios and so use leaner fuel mixtures, thus improving engine efficiency. Octane rating increases with increasing amounts of water for neat ethanol fuel [112]. Estimates for the increase in volumetric ethanol consumption relative to gasoline consumption, based on the same work done, are 10–20% [115] and 15–25% [116] for hydrous ethanol, and 25% for anhydrous ethanol [112]. The 10–25% greater volumetric consumption and 35% lower energy density of hydrous ethanol relative to gasoline imply a 23–40% higher energy yield for the ethanol fuel[1]. At an intermediate value of 17.5% greater volumetric consumption the increase in energy yield is 31%.

The higher energy yield of ethanol compared to gasoline has an important implication for consideration of replacing gasoline by ethanol. That is, the ratio of gasoline displaced per unit ethanol is greater than 1, and approximately 1.3. Thus to replace the 1982 U.S. annual gasoline usage of about 1.2×10^{13} MJ would require about 9.2×10^{12} MJ of ethanol.

2.2.4 Energetic Considerations

The energy balance for ethanol production is of most relevance for the use of ethanol as a fuel. A major point of contention in the debate over the efficacy of fuel ethanol production from starch- and sugar-rich agricultural crops has been whether net production of useful energy accompanies such processes. This subject is dealt with in detail, and with differing conclusions, in papers by Sama [117], Yorifuji [118], Johnson [119], and Parisi [114], as well as earlier papers.

Studies considering cellulosic substrates are generally in agreement that the energy balance for ethanol production from feedstocks is favorable, and more so than for ethanol production from starch or sugar crops [62, 114, 120]. Several studies [62, 70, 121] including this one (Sects. 3.2.3, 3.3) have concluded that essentially all of the at-plant energy requirements, including electricity, for producing ethanol from lignocellulosic substrates such as wood and agricultural wastes are available from burning process residues, notably lignin. If ethanol production from biomass were to occur on a scale sufficient to make a significant contribution to total transportation energy requirements, it is very likely that markets for lignin-derived aromatic chemicals would be saturated and most residual lignin would be burned to provide process heat requirements and electricity. Depending on the availability of substitute fuels, lignin might well be more valuable as a feedstock than as a process fuel for ethanol production on a smaller scale [122, 123, 124].

The energy requirements for harvesting, chipping and transporting cellulosic materials prior to processing must also be considered. Table 4 presents data for these off-site operations for a self-planted, unfertilized fuel wood case, and for a high-productivity, short-rotation, fertilized wood fuel crop. The energy expended on these

---

[1] The calculation is as follows: $(\text{work/volume})_{\text{ethanol}}/(\text{work/volume})_{\text{gas}} = 0.800$ to 0.909; $(\text{volume/energy})_{\text{ethanol}}/(\text{volume/energy})_{\text{gas}} = 1.54$. The product of these two terms has units of $(\text{work/energy})_{\text{ethanol}}/(\text{work/energy})_{\text{gas}}$ and is equal to 1.23 to 1.40.

Table 4. Energy requirements for growing, harvesting, chipping, and transporting fuel wood (MJ t^{-1})[a]

Item	Self-planted, no fertilizer (shipped as chips)[b]	High-productivity, short-rotation fertilized wood fuel crop[c]
Silviculture		640.7
Harvest	67.1	71.9
Prep. + chipping	166.9	140.0
Hauling[d]	182.2	182.2
Other	56.6	56.6
Total	472.8	1091.4 (919.3)
% Ethanol energy, only hexan used[e]	8.6	19.8 (16.7)
% Ethanol energy, hexan + pentan used	6.4	14.8 (12.5)

[a] All values include the energy devoted to equipment manufacture and maintenance as well as listed operations;
[b] data from [61];
[c] data from [61]; data in parenthesis from [62]; adding in the same transportation allowance used in [61];
[d] for a 50 mile average radius, the same value used in [51];
[e] the composition of hardwood in Table 2 is used, along with 90% substrate utilization, and 47% ethanol yield

operations relative to the energy in the ethanol which could be produced from the transported substrate depends on assumptions about the fraction of substrate which may be converted. If only hexan is used, 8.6% of the ethanol energy is used in the self-planted case, and 16.7–19.8% are used in the high-productivity case. If both hexan and pentan are used, these values become 6.4% and 12.5–14.8%.

Parisi [114] has considered energy requirements for lignocellulosic substrates other than wood. The only significant raw material-related energy requirements for materials not specifically grown for ethanol production (e.g. agricultural residues and municipal solid waste) are for collection and transportation. In the case of MSW, these can be practically zero. Transportation energy requirements for agricultural residues such as straw and corn stover are approximately 1.8 times those for wood (see Table 4).

In contrast to starch- and sugar-rich agricultural crops, the situation with respect to the "energy balance" for ethanol production from lignocellulosic substrates appears to be quite clear cut: Processing energy requirements including electricity can be met or nearly met using substrate-derived process residues with little or no supplemental biomass and no supplemental liquid fuel. Pre-processing energy requirements represent a liquid fuel requirement on the order of 10% of the energy value of the ethanol.

2.2.5 End Use

Ethanol production may alternatively be viewed as either a process yielding a chemical feedstock, a transportation fuel (for either blending or use neat), or a waste treatment process with ethanol as a valuable by-product. The level of conceptual acceptance for these uses of ethanol is probably in the order given.

Production of ethanol from biomass for use as a chemical feedstock has been favored over use as a fuel because of the differences in the amount of materials for these two uses, and consideration of the available biomass supply (see Sect. 2.1.2). Modification of the U.S. chemical industry to accomodate biologically-produced materials instead of traditional petroleum-derived feedstocks has been considered in many studies [26,27,28,56,122–126], with ethanol typically playing a key role.

In addition to the match of material demands and available biomass, the value of ethanol used for chemical and fuel applications must be considered. According to Greek [127], the alternative fuel value of any hydrocarbon is its lowest value as a chemical feedstock. This general statement appears to apply in the case of ethanol, as shown below. On a volumetric basis, and considering only energetic aspects, ethanol used as a neat fuel should be worth approximately 0.85 × the price of gasoline (based on 17.5% greater volumetric consumption, see Sect. 2.2.3). The higher octane rating of ethanol relative to gasoline can be taken advantage of in ethanol-gasoline blends by either leaving gas quality unchanged and increasing engine compression ratios and efficiency, or by decreasing gas quality to obtain a fuel suitable for use in gasoline engines [31]. Considering the latter option, the OTA [31] estimates that the value of ethanol on a volumetric basis when blended with gasoline is 1.7 times the crude oil price. Estimates for the refinery energy savings per gallon of ethanol blended 10% in gasoline vary widely from 8 to 63 MJ [31]. An intermediate value for refinery savings considering the range of estimates is 0.4 L crude oil per L ethanol, or 0.25 L gasoline equivalent per L ethanol [31]. Using this value and equivalent volumetric milage for a 10% ethanol/gasoline blend (Sect. 3.2.3) the value of ethanol as a gasoline extender is 1.25 times the value of gasoline. If a wholesale gasoline price of 0.21 $ L^{-1} is assumed, then the value of ethanol as a neat fuel is about 0.18 $ L^{-1}, and the value as an octane enhancer is 0.26 $ L^{-1}. These values may be compared to the January 1987 price for synthetically-produced ethanol, 0.44 $ L^{-1}, which reflects its value as a solvent and chemical intermediate. This simplistic analysis makes the *order* in which biomass-derived ethanol will achieve market penetration in the absence of tax incentives seem clear: chemicals before octane enhancer before neat fuel.

To consider ethanol production as a waste treatment process, the substrate must be one which is a waste disposal problem. While this is generally not the case for agricultural or logging residues, it is for municipal solid waste (MSW). Economics for alternative methods of disposal for solid wastes are highly site-specific. Recent trends in the U.S. include rapidly increasing landfill costs, increased acceptance and utilization of source separation and recycling, and consideration of electricity generation via incineration. Obviously separation of the degradable fraction raises the cost of MSW over the negative cost of tipping the unseparated waste. However, even at the price presently paid for separated recyclable paper and cardboard, highly variable but generally about 15–30 $ t^{-1}, the carbohydrate fraction of MSW appears to be a very competitive substrate for ethanol production compared to, for example, hardwood chips. Moreover, the price paid for separated paper and cardboard waste does not reflect the avoided costs of disposing of these materials.

The potential yield of ethanol from MSW is estimated at 318 L t^{-1} based on 60% degradable carbohydrate content (Table 2), 90% carbohydrate utilization, and 47% bioconversion yield. This represents a 25.4% yield by weight, and a 60.6% energy yield based on a total organic fraction (degradable and non-degradable) of 80% with a mean heating value of 13,900 kJ kg^{-1} organics. Electricity generation would be likely to have an overall efficiency of $\leqq 30\%$ based on the total organic fraction given the moisture content of MSW. The values of ethanol and electricity are similar on a per unit energy basis at approximately $ 16 per 10^6 kJ. Thus from energetic and product value considerations, ethanol production appears to be an attractive alternative to incineration and electricity generation.

3 Potential of Thermophilic Bacteria for Ethanol Production

3.1 Identification of Distinguishing Features

Table 5 presents several features of thermophilic bacteria which have been cited as either advantages or disadvantages in the literature. Reference is made to both general features and features discussed primarily with respect to ethanol production. When considered in the context of ethanol production from lignocellulose, some of these factors are of more importance than others, and some factors do not apply. It is clear that thermophilic ethanol production must compete with yeast-based processes in order to find significant utilization, thus comparison is made primarily with ethanol production using yeast in the following discussion.

By far the most important of the advantages of thermophilic bacteria listed in Table 5 for ethanol production from lignocellulose is advantage 1, the wide range of substrates utilized. This general characteristic can be divided into two specific features possessed by thermophiles collectively, but not by any single described species:

Table 5. Advantages and disadvantages of thermophilic bacteria as biocatalysts

	Refs.
Advantages	
1 Wide range of substrates metabolized (particularly pentoses and insoluble carbohydrates)	4, 5, 8, 9, 59, 128)
2 Benificial physical properties of growth medium at high temperatures (reduced viscosity and surface tension, increased diffusion rates and substrate solubility)	4, 5, 8, 9, 128)
3 High reaction rates	4, 5, 8, 9, 128)
4 Economical bioreactor cooling	4, 5, 8, 9, 128)
5 Low risk of contamination and growth of pathogens	4, 5, 8, 9, 59, 128)
6 Production of thermostable enzymes	4, 5, 8, 9, 128)
7 Low cell yields and high product yields per unit substrate	8, 9, 59, 128)
8 Low oxygen solubility	8, 9)
9 Increased value of metabolic heat	9)
10 Facilitated product recovery (primarily due to high vapor pressure of volatile compounds)	4, 5, 8, 9, 59, 128)
Disadvantages	
1 Low ethanol tolerance	8, 59)
2 Production of organic acids in addition to ethanol	8, 59)
3 Lower substrate tolerance	8)
4 Greater stress on biotechnological hardware	8, 129)
5 Sensitivity to inhibitors	130)
6 Complex growth factors required in relatively high amounts	131)
7 Limited fundamental knowledge of physiology and biochemistry	8, 9, 59, 128)

pentose utilization and cellulase production. Pentose sugars represent approximately one-fourth of the total fermentable carbohydrate in lignocellulosic substrates (Sect. 2.1.2). In light of the dominant place of substrate costs in overall process economics (Sects. 2.2.2), and that many substrate-related costs are the same with and without pentose utilization. e.g. growth, harvest, transportation, and pretreatment, the importance of pentose utilization is clear. The high cost of cellulase production and enzymatic hydrolysis for ethanol production from wood using yeast has already been alluded to (Sect. 2.2.2), and will be demonstrated further in Sect. 3.2.3.

Beneficial physical properties of growth medium at high temperatures, can be expected to allow reduced energy requirements for mixing, and perhaps other small advantages, but are unlikely to have a significant effect on overall production costs.

Advantage 3, high reaction rates, may not apply to a significant extent in the case of ethanol production. Mesophilic bacteria hydrolyze cellulose at approximately the same rate as the thermophile *Clostridium thermocellum* [132, 133]. Specific growth rates reported for thermophiles such as *Clostridium thermohydrosulfuricum*, *C. thermocellum*, *C. thermosaccharolyticum*, and *Thermoanaerobacter ethanolicus* on soluble substrates are generally between 0.4 and 0.6 h^{-1} [8,128], which may be compared to 0.43 h^{-1} to 0.46 h^{-1} for the yeast *Saccharomyces cerevisiae* growing on glucose [85, 134]. The specific rates of ethanol production by thermophiles are not substantially higher than those for either yeast of the bacterium *Z. mobilis* [6, 8]. In a more general consideration of the growth rates of various thermo-classes of bacteria, Sonnleitner and Fiechter [6] concluded that maximum growth rates for thermophiles can in general be expected to be higher than for a comparable mesophilic organism. The Arrhenius activation energy relating the rates of comparable organisms from different thermo-classes is however smaller than for a single organism. It is not clear whether the difference between this point of view and that suggested by the available data on thermophilic and mesophilic ethanol production is due to a peculiarity of ethanol production and/or cellulose hydrolysis, or to incomplete understanding of thermophilic ethanol producing bacteria.

Bioreactor cooling is not a significant cost in the production of ethanol using yeast [39, 51], and thus advantage 4 is not of great importance.

Low risk of contamination and growth of pathogens, has also been cited as a feature of yeast-based processes [8, 59], because of low pH and high ethanol concentration. The susceptability of thermophilic and yeast-based ethanol production to contamination has not been compared to the author's knowledge.

Thermostable enzymes could probably be a valuable by-product of thermophilic ethanol production, especially on a small scale. For example, thermostable starch-hydrolyzing enzymes could find extensive use in the manufacture of corn-derived sweeteners (reviewed in [10]). In the case of very large-scale ethanol production, e.g. to meet fuel demands, the demand for thermostable enzymes is very likely to become insignificant.

Low cell yields and high product yields based on higher cell maintenance requirements in thermophiles have been proposed, but not proven for thermophiles generally [5]. In a detailed study of *C. thermohydrosulfuricum*, Lacis and Lawford [135] concluded that there was no significant difference between the energy required for cellular maintenance in this thermophile and values for this parameter generally reported for mesophiles. To the author's knowledge, there is no correspondingly strong evidence that favors the interpretation that thermophiles have higher maintenance requirements. In any case, the cell yield is clearly of secondary importance compared to the synthesis of organic acids, discussed below, in affecting the ethanol yields of thermophilic processes.

At the scale and cell concentrations required for a practical process it is very unlikely that the lower oxygen solubility at elevated temperatures, advantage 8, would be of any consequence.

Calorimetric studies of cellulose utilization by anaerobic bacteria have reported the production of metabolic heat at a level which corresponds to about 1.27 MJ L^{-1} [136], assuming ethanol is the only soluble product formed. This quantity of heat, assuming that it all is recovered, represents approximately 1/4 of distillation energy requirements and 1/6 of total energy requirements for production of ethanol from corn using yeast; total energy requirements for ethanol production from corn using yeast account for about 13% of production costs [37]. Thus the potential economic impact of metabolic heat is likely to be on the order of a few percent of production costs. Use of metabolic heat is restricted to process heat requirements at temperatures less than the temperature of bioconversion. These include preheat of various process streams, and also product recovery in cases where this step is carried out at suitably low temperatures, e.g. reduced pressure distillation. In some situations, the contribution of metabolic heat may be significant, and should not be discounted. At the least, metabolic heat can contribute to maintaining the bioreactor at thermophilic temperatures, for which a counter-current heat exchanger heating the bioreactor feed is also helpful.

Facilitated product recovery is often cited as an advantage of thermophilic bacteria for ethanol production (see Table 5). This factor will be an advantage of thermophilic systems compared to mesophilic systems at the same ethanol concentration, and a distinct advantage for thermophilic systems if continuous ethanol removal is employed. However, it is the author's evaluation that facilitated product recovery has been over-rated as an advantage of thermophilic systems, particularly when comparing thermophiles to yeast. Contrary to some reports [137], there is no particular significance to the fact that some thermophilic ethanol producers will grow at temperatures near the boiling point of ethanol, since aqueous solutions of a few % ethanol boil at >95 °C. It is true that operation at a higher temperature may decrease the temperature difference between the fermentor and distillation system, which is normally operated at ≧ atmospheric pressure. The result of this may be only to change the size of the preheat heat exchanger since many process streams, such as the stripping section bottoms, are often available to provide preheat. At most the preheat energy requirement will be saved. The energy required for preheat can be significant at the low ethanol concentrations processes using thermophiles may have to operate at, but this energy requirement is a small fraction of distillation energy requirements for processes using yeast. It is also true that the pressure for boiling spent medium at the conversion temperature is higher for thermophilic systems than mesophilic systems, but the pressures are still low in most cases. For example the pressure must be ~0.2 atm to boil an ethanol-water mixture at a few % ethanol at 60 °C. This may be compared to ~0.09 atm for operation at 37 °C. Because of the higher pressures which may be used, continuous ethanol removal at thermophilic conversion temperatures should involve lower capital costs, which is fortunate because ethanol removal may be necessary to obtain reasonable productivites. However energy is not saved by virture of the difference in pressures for continuous ethanol removal. The latent heats are slightly larger at low pressure and thermophilic temperatures, and the relative volatility of ethanol is slightly lower than at normal distillation temperatures both negative factors for distillation energy requirements. The "gas stripping ethanol rectifier" discussed by Slapack et al. [8], and also other similar gas stripping processes for continuously removing ethanol during thermophilic ethanol production [6,7], reduce subsequent distillation energy requirements, as claimed, but will generally have an unacceptably large energy requirement for vapor evaporation in the gas stripping step[1].

An important disadvantage of thermophilic bacteria relative to yeast is low ethanol tolerance. Low ethanol tolerance is undesirable because it increases the cost of product recovery. In addition, low ethanol tolerance limits allowable substrate concentrations, thus decreasing productivity and increasing bioreactor costs. As pointed out by Maiorella et al. [85], low feed substrate concentrations also have economic penalties other than bioreactor costs, such as costs related to sterilization, evaporation, drying, and waste treatment.

In general the available data appears to be consistent with the assumption of similar ethanol production rates as a function of substrate and cell concentrations for both yeast and thermophiles using soluble substrates. It may also be reasonable to assume that a similar function describes the degree of ethanol inhibition in relation to the maximum ethanol concentration for both systems. For kinetically identical systems, volumetric bioreactor productivity will be directly related to the maximum ethanol concentration tolerated. Yeast and thermophilic bacteria appear to differ by a factor of approximately 3 with respect to this key parameter (see Sect. 4.3). Thus volumetric productivity for thermophilic and yeast-based ethanol production systems in the same kind of bioreactor may be expected to differ by about 3-fold. Volumetric productivity for in situ thermophilic lignocellulose utilization is likely to be less than for thermophilic utilization of soluble substrates by a further factor of 3 due to lower growth rates.

With the exception of ethanol recovery costs, the other limitations and costs associated with low ethanol tolerance can in principle be aleviated by continuously removing ethanol from the bioreactor, and so uncoupling substrate and product concentrations. The incentive to utilize continuous ethanol removal with thermophilic ethanol production systems appears to be strong, especially since a favorable impact on ethanol yields may also be achieved (see Sect. 4.4). Economically- and physiologically-acceptable methods of continuous ethanol removal have not been demonstrated for thermophilic systems even on a laboratory scale (see Sect. 5).

The fraction of substrate converted to ethanol, which is decreased by organic acid production, is a key economic factor because of the dominance of substrate costs.

---

[1] For example: Ethanol and water have essentially equal latent heats on a molar basis at about 40.6 kJ g mole^{-1}. If an inert gas is bubbled through a 2 wt. % ethanol aqueous solution (liquid ethanol mole fraction, Xethanol, = 0.00792) to continuously remove ethanol, the liquid which may be obtained by condensing the inert gas/ethanol/water mixture will have mole fraction $X_{ethanol} = 0.086$, with the remainder water. To recover 1 mole of ethanol thus requires evaporation of 11.6 moles of liquid, corresponding to a heat requirement of 472 kJ g mol^{-1} ethanol. This heat requirement represents about 38% of the combustion energy of the ethanol (1231 kJ g mole^{-1}), and subsequent distillation to complete ethanol separation will require further heat.

Engineering aspects of ethanol tolerance and yield will be discussed further in Sect. 3.2.3, biological aspects will be discussed in Sects. 4.3 and 4.4.

Low substrate tolerance would not be expected to be a factor in continuous culture, though it does limit the utility of the batch processes. In addition to substrate tolerance, other factors favoring continuous systems over batch systems for thermophilic ethanol production are the generally higher productivity of continuous systems [37, 84], and the relative ease of ethanol removal.

Sensitive to inhibitors, particularly arising from pretreatment, and also a possible requirement for more elaborate and costly growth medium, could be significant limitations for thermophilic ethanol production if they were shown to be widespread fundamentally-based characteristics. However, optimization of medium formulation and substrate pretreatment to minimize by-product inhibition are in very preliminary stages with respect to thermophilic systems. Moreover, similar problems have been solved for other systems. Thus it is a distinct possibility that disadvantages 5 and 6, along with others, may be mitigated by addressing disadvantage 7, limited fundamental knowledge of the physiology and biochemistry of thermophilic ethanol-producing bacteria.

In summary, it is the author's evaluation that in comparison to ethanol production using yeast, the two most important advantages of thermophilic bacteria for ethanol production from lignocellulosic materials are pentose utilization and cellulase production. The two most important disadvantages are low ethanol tolerance and organic acid production.

An interesting observation is that neither pentose utilization nor cellulase production are associated with growth at high temperatures per se. Thus obtaining non-thermophilic organisms with these properties is an endeavor with some justification. Several studies have demonstrated some degree of expression of genes coding for one or more cellulase components from mesophilic organisms in a convenient host [138, 139, 140], and preliminary studies have also been carried out on cellulase expression in yeasts [141], and on gene transfer systems for *Zymomonas mobilis* [142]. However, obtaining coordinated expression of foreign genes coding for synthesis and secretion of a multi-enzyme complex, together with transport and catabolism of resulting soluble substates is an ambitious undertaking. By contrast the achievements of genetic enginering to date typically involve expression, often without secretion, of a single gene [143]. A second observation about the advantages of thermophilic bacteria for ethanol production is that neither pentose utilization nor cellulase production is a significant factor for starch and/or hexose-rich substrates such as corn and juice from sugar cane. In contrast to cellulase, amylase is relatively inexpensive to produce [39].

3.2 Evaluation of Distinguishing Features

Having identified the important distinguishing features of thermophilic bacteria for ethanol production from lignocellulose, it is desirable to evaluate these features. This task is made more difficult by the fact that there have to date been no experimental studies of thermophilic bacteria with an economically acceptable combination of substrate and medium composition, bioreactor productivity, ethanol concentration, and ethanol yield. Progress toward realization of the goal of economic ethanol production using thermophilic bacteria is considered in Sect. 4.

The approach taken here is to consider a detailed process design and economic analysis for ethanol production from lignocellulose via enzymatic hydrolysis and yeast, and then to present the economic impact of process changes associated with the distinguishing features of thermophiles. The purpose of this exercise is to determine if and to what extent these features impact strategic cost factors in the context of overall process economics. In so doing the relative importance of the distinguishing features of thermophiles and the potential economics of thermophile-based and other ethanol production processes can be compared.

A detailed plant design and economic evaluation of an enzymatic hydrolysis-based plant producing 94.6×10^6 L a^{-1} of ethanol via yeast is used as a base case. This design, by Chem Systems Inc., Tarrytown NY [51], utilizes dilute acid hydrolysis for substrate pretreatment, a process which has been shown to be effective in allowing high extents of hydrolysis for both in vivo and in vitro thermophilic systems (Sect. 4.1). By-product credits are taken for both furfural, produced from xylose and valued here at 33 cents kg^{-1}, and CO_2, valued at 4.4 cents kg^{-1}. High pressure steam is expanded in a turbo-generator to produce electricity, and the exhaust steam is used for process heat requirements. Process residues and unprocessed wood amounting to 11% of the total wood used provide all process heat and 95% of required electricity. Raw materials required in addition to the primary substrate include medium chemicals, H_2SO_4, lime, and cellobiase.

3.2.1 General Impact

The primary effects of the positive distinguishing features of thermophilic bacteria, cellulase production and pentose utilization, are process simplification and an increase in the fraction of substrate which could be converted to ethanol. A schematic flow sheet of the Chem Systems design for ethanol production from wood is shown in Fig. 4a. Important process steps include pretreatment, separation and neutralization, enzyme production, enzyme hydrolysis, sugar concentration, biological conversion to ethanol, purification, furfural production, heat generation, waste treatment and CO_2 recovery. Fig. 4b presents a flow sheet for a hypothetical thermophilic process assuming that cellulase production and substrate hydrolysis occur in the fermentor, and 5-carbon as well as 6-carbon sugars are converted to ethanol. It is evident that there are fewer process steps and fewer divided flows, the latter often representing costly solid-liquid separations.

The amount of degradable carbohydrate available from a given amount of wood leaving the pretreatment section will be greater for the thermophilic design (Fig. 4b) than for the mesophilic design (Fig. 4a). This difference is due to 2 factors: 1) that no solids and accompanying soluble sugars are diverted to a separate cellulase production step, and 2) that pentoses are used in the thermophilic case. The substrate flow in the base-case design is presented in Fig. 5. All values represent fractions of material per unit of original substrate using the substrate composition assumed for the base case. Carbohydrate fractions are calculated on a monomer basis, that is including the weight added by the water of hydrolysis, throughout. The yield implications of various process modifications are presented in Table 6. It may be seen that

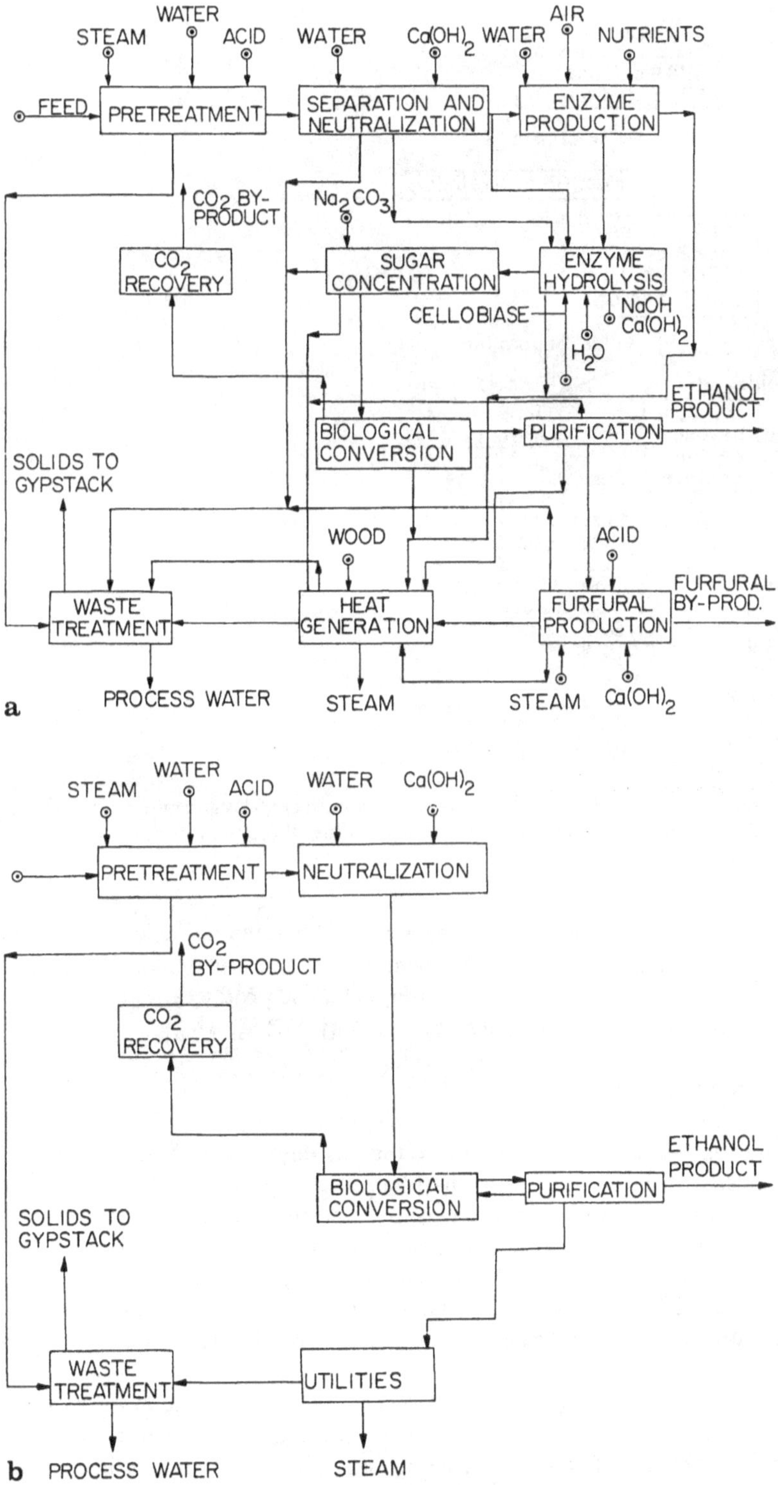

Fig. 4. Comparison of flow sheets for ethanol production from wood using yeast and enzymatic hydrolysis, Fig. 4a, and thermophilic bacteria, Fig. 4b. Fig. 4a is essentially as presented in [51]

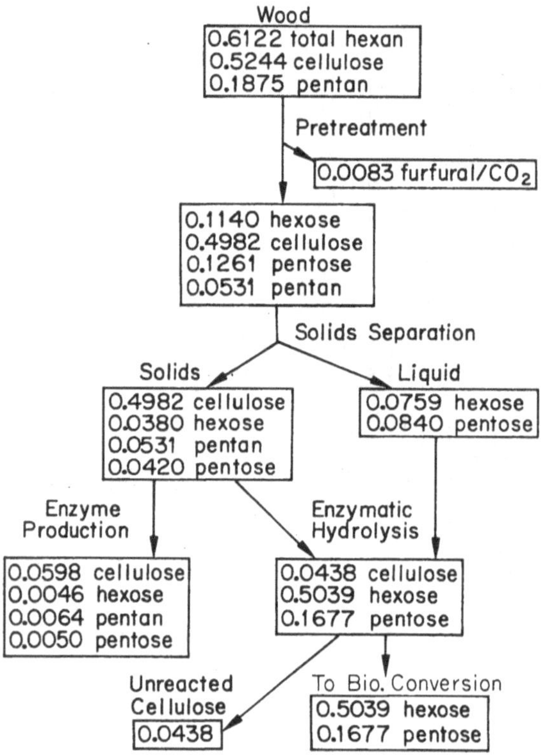

Fig. 5. Substrate flow in the Chem Systems [51] wood to ethanol process design employing yeast and enzymatic hydrolysis. All values are on a monomer basis (adjusted throughout for the water of hydrolysis) per unit substrate as originally received

eliminating the cellulase production step increases the degradable carbohydrate per unit substrate by 11.6%, utilization of pentoses (while still making cellulase) results in a 33.3% increase, and both eliminating cellulase production and utilizing pentoses increases the degradable carbohydrate yield by 47.1%.

3.2.2 Basis for Economic Analysis

In modifying the base-case process to reflect the distinguishing characteristics of thermophiles, the size of the equipment for many process steps are changed. A power law is used to adjust the cost of this equipment in relation to changes in size. Power law exponents are derived from costs given by Badger [50] for ethanol plants with different production capacities (the Chem Systems study did not give alternative capacities). Raw materials, by-product credits, and utility-related operating costs are assumed to be proportional to the rate of utilization or production, as in the Chem Systems study.

Several process steps perform the same functions in the base case and in the thermophilic case. These include wood handling, pretreatment, heat generation, waste treatment, and CO_2 recovery. These steps are only changed due to changes in capacity

Table 6. Fate of carbohydrate species in the chem systems plant design for ethanol production from wood using yeast and enzymatic hydrolysis

Hexose:	
0.6122	0.5039 to bioreactor (H_f)
	0.0598 cellulose to cellulase production ($H_{c,c}$)
	0.0046 hexose to cellulase production ($H_{c,h}$)
	0.0438 to unreacted cellulose (H_u)
	0.6121
Pentose:	
0.1875	0.1677 to bioreactor (P_f)
	0.0064 pentan to cellulase production ($P_{c,n}$)
	0.0050 pentose to cellulase production ($P_{c,s}$)
	0.0083 to furfural/CO_2 (Pd)
	0.1874

Yield implications of process modifications

I. Eliminate cellulase production:

$$\text{Yield multiplyer (Y.M)} = \frac{H_f + H_{c,c} \times 0.9 + H_{c,h}}{H_f} = 1.116$$

II. Use pentoses:

$$\text{Y.M.} = \frac{H_f + P_f}{H_f} = 1.333$$

III. I and II:

$$\text{Y.M.} = \frac{H_f + H_{c,c} \times 0.9 + H_{c,h} + P_f + P_{c,n} + P_{c,s}}{H_f} = 1.471$$

All values are on a monomer basis (adjusted throughout for the water of hydrolysis) per unit substrate as originally received

Values from Fig. 5 as given by [51]

in the thermophilic case. Changed cost due to capacity only is also assumed for the biological conversion section. A bioreactor designed for thermophilic ethanol production would probably be somewhat different from the types used in the base case because solids are separated prior to bioconversion and are used directly when thermophiles are used, and also because of other factors. However, the base-case bioreactor system has a residence time of 18 h, which is similar to but somewhat greater than the 12 to 16 h residence times required to achieve >85% utilization of pretreated mixed hardwood by *C. thermocellum* in continuous culture, as discussed in Sect. 4.1. Given these considerations, the lack of data upon which to base a design for the thermophilic system, and the generally non-critical nature of bioreactor costs, the cost of the biological conversion section is assumed to be the same, with adjustments for changes in capacity, for the thermophilic case and for the base case. The bioreactor is operated at atmospheric pressure in both cases.

In this study a bioreactor ethanol concentration of 1.5% is assumed. This value has consistently been produced by both batch and continuous thermophilic cultures (see Sect. 4.3 for further discussion of ethanol tolerance). Liquor from the bioreactor is allowed to flow to the distillation system, stripped, and pumped back to the fermentor, all at 60 °C. Low ethanol concentrations are thus maintained in the bioreactor though the substrate concentration leaving the pretreatment reactor is 16.5 wt. % solids in both the base and thermophilic cases.

Conventional ethanol distillation is very unlikely to be satisfactory for separating ethanol at 1.5 wt. % because of high steam requirements. Therefore the IHOSR/extractive distillation process is used (see Sect. 2.2.2). The size and cost of equipment for this process is based wherever possible on the design parameters and costs of similar equipment in the Chem Sytems design. Standard cost correlations and costing procedures are used [144, 145].

The Chem Systems study assumes 90% enzymatic hydrolysis yield using cellulase from *Trichoderma reesei* acting on a wood feed consisting of 57% aspen, 20% maple, and 23% other woods after pretreatment at 200 °C for 12 s with 0.75% H_2SO_4. Data are presented in Sect. 4.1 which show that in vitro and in vivo yields from hydrolysis using *Clostridium thermocellum* cellulase are consistently comparable to hydrolysis yields using *T. reesei* cellulase. Thus the same hydrolysis yield is used in the thermophilic case as in the base case.

A major assumption is that high substrate concentrations can be converted by the thermophilic system with the same utilization of degradable carbohydrate observed at low substrate concentrations. While this would be expected given sufficient knowledge of nutritional requirements, it has not been demonstrated. Also, no change is made in the cost of chemicals for medium supplements (nitrogen source etc.) compared to the base case. Finally, it is assumed that there is no inhibition of growth or hydrolysis by pretreatment by-products. The issues of utilization of high substrate concentrations, nutrient supplement requirements, and by-product inhibition are some of the perhaps less glamorous research areas which must be addressed in order to develop thermophilic ethanol production. Making optimistic assumptions about these issues is based on the fact that similar issues have been solved for other systems. It is also consistent with the goal of this analysis, that is, to investigate the economic implications of distinguishing features of thermophiles as biocatalysts for ethanol production.

Given a process with higher yields per unit substrate, one could either keep the ethanol production the same and reduce the substrate flow, or keep the substrate flow the same and increase the output. Since the major issues in determining plant capacity are related to the substrate (e.g. the available substrate supply and the distance it must be transported) and not the product, it is elected to keep the flow of substrate to the pretreatment reactor the same as in the base case, and to let the ethanol production increase. This actually results in a ~11% smaller total wood usage since the base case design uses some wood fed directly to the boiler whereas the thermophilic design does not.

3.2.3 Economic Impact

A detailed consideration of the economic impact of the distinguishing features of thermophiles for ethanol production from lignocellulose has recently been completed by the author [146]. A summary of the results is presented here.

Table 7 presents capital costs for production of ethanol from wood using yeast and enzymatic hydrolysis, the base case, as reported by Chem Systems [51]. Also presented in Table 7 is the thermophilic case representing the cumulative economic impact of three of the four distinguishing features of thermophilic bacteria: pentose utilization, cellulase production, and low ethanol tolerance. Production of organic acids is not

Table 7. Capital costs for production of ethanol from wood using yeast and enzymatic hydrolysis, and the capital cost-related impact of cellulase production, pentose utilization, and operation at 1.5 wt. % ethanol[a]

Item	U.S.$ × 10^{-6}	
	Base case	Thermophilic case
1 Wood handling	8.10	7.74
2 Pretreatment	3.10	3.10
3 Sugar separation/neutralization	3.90	0.72
4 Cellulase production	6.38	0
5 Enzymatic hydrolysis	7.07	2.94
6 Sugar concentration	5.94	0
7 Biological conversion	4.21	4.87
8 Ethanol recovery	4.11	11.28
9 Utilities	28.86	25.29
10 By-product processing	10.97	6.73
11 Storage	1.93	1.68
12 Pollution control	2.86	2.58
Total direct	87.43	66.93
Indirect	24.58	15.00
Total fixed investment	112.0	81.93

[a] Base case in mid 1984 $ as presented by Chem System [51], for a plant producing 94.6×10^6 L a^{-1} ethanol. The thermophilic case is in mid 1984 $ for a plant processing the same amount of wood through the pretreatment section and producing 139×10^6 L a^{-1} ethanol. See text and [146] for further details

considered in Table 7. The reasons for the differences in the capital costs for items in the base and thermophilic cases are discussed below.

In situ cellulase production eliminates or greatly reduces costs for items 3, 4, 5, and 6, all of which are devoted to either production or efficient utilization of cellulase in the base case. Costs for certain functions accomplished in these items are retained, in particular for solids concentration. Solids concentration is accomplished in step 5 in the base case, and is accomplished following biological conversion and prior to burning process residues in the thermophilic case. Pentose utilization eliminates furfural production and processing, a part of item 10, and associated storage and waste treatment in items 11 and 12. Pentose utilization in combination with cellulase production increases the ethanol output by 47% from 94.6×10^6 L a^{-1} to 143×10^6 L a^{-1}, as discussed in Sect. 3.1, with consequent increases in costs for biological conversion and CO_2 recovery. The cost of distillation, item 8, is increased by both the increased ethanol output and also due to the use of IHOSR/extractive distillation to recover ethanol at a concentration of 1.5 wt. %. Both the effect of cellulase production in eliminating portions of items 3 through 6, and also the substitution of the more efficient IHOSR/extractive distillation for the conventional distillation used in the base case contribute to reducing utility-related capital costs, item 9, in the thermophilic case. Of the costs for distillation in the modified thermophilic case, 73% are for two kinds of items: eight 10ft diameter disc and doughnut-type reduced pressure stripping columns, and a 3400 horsepower compressor which drives the heat pump cycle. Compressor technology is relatively well-developed, and costs and performance may be predicted with relative certainty. There is more uncertainty about the cost and performance of the stripping columns, principally because the influence of cells and substrate residue on stripping efficiency and the VLE relationship have not been adequately studied.

Table 8 presents production costs for the base case and the thermophilic case. Wood costs are different because excess wood is required to meet process steam requirements in the base case, but not in the thermophilic case. Acid and lime requirements are lower in the thermophilic case because furfural production is eliminated. Utility-related operating costs are lower for the same reasons capital costs for utilities are lower. 95% of electricity requirements are generated on-site in the base case, and 93% in the thermophilic case. By-product credits for CO_2 are increased in the thermophilic case due to the increased carbohydrate metabolized, but are eliminated for furfural. Labor costs are lower in the thermophilic base case because they are more closely related to the number of process steps than to production capacity. Maintenance and taxes are related to capital costs, overhead is related to labor and capital costs.

A capital recovery factor based on a 15% before taxes return on investment over 10 years for the case of 100% equity financing is used in Table 8 to convert capital

Table 8. Production costs for producing ethanol from wood using yeast and enzymatic hydrolysis, and the production cost-related impact of cellulase production, pentose utilization, and operation at 1.5 wt. % ethanol[a]

Item	U.S.$ × 10^{-6}	
	Base case	Thermophilic case
Raw materials		
Wood	15.06	13.39
Acid	1.56	0.86
Lime	0.67	0.36
Cellobiase	1.35	0
Medium components	1.61	2.37
By-product credits		
Furfural (33 cts kg^{-1})	−5.49	0
CO_2 (6.2 cts kg^{-1})	−4.56	−6.71
Operating costs		
Labor	1.69	1.26
Maintenance	3.98	2.58
Purchased power	0.34	0.40
Labor, maintenance and[2] *overhead for utilities*		
Generated power	1.29	1.10
Boiler feed water	0.22	0.15
Cooling water	1.07	0.75
Steam	2.05	1.50
Overhead and taxes	6.12	4.29
Total production costs	26.96	22.30
Capital recovery (15% R.O.I, 10 years)	22.32	16.33
Ethanol production (L a^{-1})	94.6	139
Selling price (U.S.$ L^{-1})	52.1	27.8

[a] All costs in mid 1984 dollars. Base case and thermophilic case as for Table 6;
[b] labor, overhead, and maintenance are charged separately for utilities, which are considered off-site, and process equipment, which is considered on-site

costs into an annual operating cost. Using this procedure, the selling prices for ethanol, including return on capital investment, are 0.52 \$ L^{-1} for the base case, and 0.28 \$ L^{-1} for the thermophilic case. All costs are in mid 1984 U.S. dollars.

Relative to the base case selling price of 0.52 \$ L^{-1}, the impact of cellulase production considered individually is to reduce the selling cost by 19 cts L^{-1} ethanol, or 37%. The impact of pentose utilization relative to the base case selling price is to reduce the selling price by 12 cts L^{-1} ethanol, or 23%. The influence of these factors on the final selling price is not additive. The contribution of distillation to the ethanol selling price, considering capital and operating cost factors including utilities is 2.9 cts L^{-1} for the base case, 5.1 cts L^{-1} for the thermophilic case at the same 94.6×10^6 L a^{-1} capacity, and 4.3 cts L^{-1} for the thermophilic case at the increased capacity as considered above. Thus the added cost of operating at 1.5% ethanol with continuous ethanol removal is relatively small.

Reductions in the ethanol selling price due to cellulase production and pentose utilization presented above are supported by economic data presented by Wright et al. [70], for a plant producing ethanol from hardwood using yeast and enzymatic hydrolysis. The design analyzed by these workers differs from the Chem Systems design considered thus far in that furfural is not produced and steam explosion is used as a pretreatment. According to Wright et al. [70], cellulase production and enzymatic hydrolysis together are responsible for approximately 40% of total ethanol production costs, and conversion of pentoses to ethanol would reduce the selling price by 34%. This value for the impact of pentose utilization is higher than that presented above, presumably reflecting the benefit of furfural production in the Chem Systems design.

Consideration of the impact of lower than theoretical ethanol yield involves an assessment of the value or disposal cost of by-products. Both acetic and lactic acids, the main products other than ethanol produced by many thermophilic bacteria, have values at least that of ethanol, and can be produced in higher yield from sugars because of their higher molecular weights. However, recovering organic acids by distillation from dilute solution is considerably more difficult then recovering ethanol because the boiling point of acetic acid is higher than water, so the water must be boiled away from the acetic acid instead of vice-versa [103]. Separation technologies other than distillation, such as solvent extraction are not established and are also costly [103]. Dupont investigated biological acetic acid production in the early 1980's and concluded it was uneconomical, largely because of separation issues [Dr. Thomas Ng, Dupont, personal communication]. Similarly, processes based on non-biological production of fuels from organic acids [147] have also been abandoned.

Both acetic and lactic acid make excellent substrates for methane production, however methane is less valuable than ethanol as both a chemical and a fuel, and is produced in 50% lower mass yield. Thus even if methane digestion had no production costs, a substantial portion of the value of ethanol would be lost for every unit of organic acids converted to methane. Methane production from soluble organics was considered in the Chem Systems design as an alternative to multieffect evaporation to concentrate the organics followed by combustion; the two options were roughly equivalent in cost.

If acetic or lactic acid is co-produced with ethanol in a wood-based plant, the economics for both products will suffer because there will be equipment specifically devoted to each product, produced on a smaller scale than if it were the only product, and the size of the plant is limited by the availability of substrate. Moreover, co-production of ethanol and organic acids is likely to face severe problems of unequal demand if ethanol is produced on a scale able to contribute significantly to fuel requirements.

In light of the considerations above, the assumption that organic acids can be disposed of with no net revenue or cost may be realistic for many cases, particularly when large-scale ethanol production is considered. In the case of large scale production, there is a very great incentive to achieve high ethanol yields in thermophilic processes. In particular, the discussion above points to the greater sensitivity of process economics to ethanol yield than to ethanol tolerance. For example an 80% decrease in ethanol concentration relative to the base case brought about a 2.2 cts L^{-1} increase in the ethanol selling price. This same increase would be expected to arise from a roughly 8% decrease in the ethanol yields.

Based on the consideration of the distinguishing features of thermophilic bacteria presented above, it is concluded that the two key advantages of thermophilic bacteria, cellulase production and pentose utilization, potentially have large impacts on overall process economics. Specifically they reduce the ethanol selling price for ethanol production from wood using enzymatic hydrolysis and yeast by about a factor of 2. Approximately the same cost reduction is realized in comparison to acid hydrolysis-based processes, since production costs are similar for acid- and enzymatically-catalyzed cellulose hydrolysis [50]. At least for the case where ethanol is continuously removed, ethanol tolerance does not appear to be as important a limitation for thermophilic ethanol production as has often been claimed. The capital costs for conventional distillation are a relatively small fraction of total capital costs in yeast-based ethanol production. Thus a process which can more efficiently separate ethanol at low concentrations can be substituted in the thermophilic case with a relatively small cost penalty. The IHOSR/extractive distillation process considered above appears to be one such process. High ethanol yields appear to be a requirement for practical thermophilic ethanol production at high production levels. At lower levels of production, significant production of organic acids could be acceptable only if progress were made in the areas of organic acid recovery and/or conversion to fuels.

3.3 Comparison with Other Ethanol Production Processes

Since wood-based processes are in the research stage and are presently uneconomical, it is of interest to compare the potential economics for ethanol production from wood using thermophilic bacteria with ethanol production processes presently in use. Table 9 compares economics for thermophilic ethanol production from wood, based on the analysis above and assuming high ethanol yields, with economics for ethanol production from corn using yeast, based on the design by Katzen et al. [39], and also for production from ethylene using data presented by Hacking [56]. For the sake of this comparison, no by-product credit for CO_2 is taken for thermophilic production from wood, and maintenance, overhead, taxes and insurance are calculated on a common basis. When CO_2 is not recovered, all process fuel and electricity requirements are satisfied from burning process residues in the thermophilic case. The ethanol selling prices presented in Table 9 for the three processes are essentially equal. This

Table 9. Comparison of the potential economics of ethanol produced from wood using thermophilic bacteria with economics for ethanol produced from corn using yeast and from ethylene by Chemical synthesis (mid 1984 U.S. $)

	I Wood/thermophilies (potential, no CO_2 recovered)[a]	II Corn/yeast[b]	III Ethylene/catalyst[c]
Capacity ($L\ a^{-1}$)	139×10^6	139×10^6	189×10^6
Capital cost ($\$ \times 10^6$)	74.4	70.6	57.8
Production cost (cts L^{-1})			
Substrate	9.62	22.88	16.57
Other raw materials	2.59	2.27	1.29
Labor, maintenance, overhead. taxes and insurance[d]	6.12	6.01	4.39
Fuel and electricity[e]	0	1.74	4.52
By-product credits	0	−13.21	0
Total production cost	18.33	19.69	26.7
Capital recovery (15% R.O.I., 10 years)	10.65	10.09	6.08
Selling price	29.0	29.8	32.9

[a] Costs are based on the modified thermophilic case considered in Tables 7 and 8, but with no recovery of CO_2;
[b] capital, labor, and utilities costs are based on the design of Katzen and Associates [39]. Adjustment was made for scale by interpolating between costs given for plants with capacities of $94.6 \times 10^6\ L\ a^{-1}$ and $189.3 \times 10^6\ L\ a^{-1}$. Adjustment is also made for the year of the cost estimates from the end of 1978 to mid 1984 using standard indices for capital and coal. A price of 0.06 \$ kWh^{-1} is used for purchased electricity as in the Chem Systems study [51]. The price of the distillers' dried grain by-product is based on data from [37]. A corn price of 3 \$ $bushel^{-1}$ (0.12 \$ kg^{-1});
[c] capital costs and materials other than ethylene are based on values presented by Hacking [56] taken from data of Flannery and Steinschneider, adjusted from first quarter 1981 to mid 1984 using a 6% per year growth rate. The price of ethylene is taken to be 33 cents kg^{-1}. Utilities are from personal communication with Genaro Maffia, Arco Oil;
[d] these items are charged for on a common basis as follows: maintenance, 4% of capital costs; overhead, 0.5 × (labor + maintenance); taxes and insurance, 2% of capital costs;
[e] costs are for fuel and electricity for I and II, and for fuel, electricity and also other utility-related costs, such as feed water conditioning, for III

suggests that if ethanol yields comparable to yeasts can be achieved by thermophiles, and other assumed characteristics of thermophiles, e.g. utilization of high substrate concentrations, are also experimentally demonstrated, then thermophilic ethanol production from wood can be expected to be competitive with yeast-based production from corn and synthetic ethanol production from ethylene. If the high yields and other assumed characteristics were realized by thermophilic systems, then features of thermophiles ofter than the distinguishing features considered thus far might become important in determining the competitiveness of the processes considered in Table 9.

The similarity of the cost structure for thermophilic ethanol production from wood and yeast-based ethanol production from corn as presented in Table 9 is striking. In particular the net cost of substrate and capital costs are very similar. Clearly the corn/yeast case is uninteresting in the absence of by-product credits. Of the three processes, the selling price of ethanol produced from ethylene is the most sensitive to raw material costs and the least influenced by capital costs. It may be noted that in 1982 the real price of ethylene was nearly twice the 33 cts kg^{-1} value assumed in Table 9. Factors such as operation at ethanol concentrations greater than 1.5% and utilization of a less expensive substrate, for example municipal solid wastes, would make the thermophilic process more competitive.

4 Progress Toward Realization of the Potential of Thermophilic Bacteria for Ethanol Production

Having identified cellulase production, pentose utilization, ethanol tolerance, and ethanol yield as the important distinguishing features of thermophiles for ethanol production (Sect. 3.1), research results in each of these areas are reviewed and evaluated.

4.1 Cellulase Production and Activity

The production of cellulase by thermophilic bacteria is one of their most important attributes for ethanol production (Sect. 3). Knowledge of cellulase production and activity in thermophilic bacteria is essentially restricted to *Clostridium thermocellum*. The cellulase activity of *C. thermocellum* is predominantly cell-associated in exponential phase, and cell-free in stationary phase [148, 149]. Glucose and cellobiose are the main products of cellulose hydrolysis [150, 151]. In addition to cellulose hydrolysis, *C. thermocellum* cellulase also catalyzes xylan hydrolysis [150]. Cellulase or other extracellular enzymes of *C. thermocellum* also catalyze the hydrolysis of pectin [152].

Johnson et al. [151] have demonstrated that the cellulase system of *C. thermocellum* catalyzes complete hydrolysis of purified crystalline cellulosic substrates such as Avicel, cotton, and filter paper. Rates of cellulose hydrolysis are comparable for reaction mixtures from *C. thermocellum* 27405 and *Trichoderma reesei* QM414. The specific activity of *C. thermocellum* cellulase is much greater than that of *T. reesei*. Hydrolysis of crystalline cellulose requires Ca^{++} and is inhibited by oxygen [151], and also by cellobiose [153]. Cellulase activity on crystalline substrates has characteristics distinct from activity on noncrystalline substrates with respect to both product inhibition [153], and thermal deactivation [151]. The formation of cellulase with activity toward crystalline substrates is repressed by cellobiose, and derepressed by Avicel [154].

At least two protein components have been found to be required for crystalline cellulase activity by Wu and Demain [155], though many more components are present [156, 2, 157]. Recently a single protein produced by *C. thermocellum* apparently with both endo- and exo-glucanase activity has been isolated [158]. A cellulose-binding, cellulase-containing complex produced by this organism has been isolated by Lamed et al. [159] which is responsible for most of the activity observed in culture broths and has properties consistent with those of unpurified extracellular protein [160]. This complex, termed a cellulosome, is composed of at least 14 distinct polypeptides with a combined molecular weight of approximately 2.1 MDa. During exponential growth on cellulose, the cellulosome is anchored to the cell surface, and allows adherence of the cell to the insoluble cellulosic substrate. The rate of cellulose hydrolysis by intact cells is significantly higher than that of the cell-free cellulase system [2].

Coughlan et al. [161] have reported that *C. thermocellum* cellulase can be resolved into 2 major complexes with diameters of 210 Å and 610 Å and molecular weights of 4.2 and 102 MDa respectively; other reports have been consistent with these very large dimensions [156, 159]. These values may be compared to a representative diameter of 51 Å [1] and component molecular weights of <100 kDa [162] for *T. reesei* cellulase. In spite of this great difference in size, cellulase produced by *C. thermocellum* is roughly as effective as cellulase produced by *T. reesei* in attacking microporous substrates which have progressively smaller surface area accessible to larger molecules [163]. Furthermore, the minimum pore dimension required for access by *C. thermocellum* cellulase is essentially the same as that required by the cellulase of *T. reesei*, about 43 Å [68]. The apparent contradiction between the reported sizes of these cellulase systems and the pore size required for substrate accessability raises questions about structure-function relationships in the cellulase enzyme complex of *C. thermocellum*.

The most important practical question regarding *C. thermocellum* cellulase is its effectiveness against complex, lignin-containing cellulosic substrates. Data are available on the cellulase activity toward lignocellulose from in vivo studies [130, 164, 165, 166, 167, 168], however the properties and limitations of the enzyme are not easily distinguished from those of the organism in these studies.

A direct in vitro evaluation of the effectiveness of *C. thermocellum* cellulase on lignocellulosic substrates is possible based on recent work by the author and collea-

Table 10. Comparison of hydrolysis yields for various systems including *C. thermocellum* or *T. reesei*

Source of cellulase	Substrate[1]	Pretreatment conditions			Solids Concentration[2]	Enzyme Loading[3]	Reaction Time[4]	Yield[5]	Refs.
		Temp. (°C)	$[H_2SO_4]$ (%)	Time (s)	(g L^{-1})	(IU g^{-1})	(h)	(%)	
A. In vitro									
C. thermocellum	Hardwood	220	1.0	9.0	2.5	36.0	13	95.0	169)
T. reesei	Hardwood	220	1.22	7.2	14.0	24.3	24	85.6	170)
C. thermocellum	White pine	220	1.0	10.0	0.52	7.0	47	43.0	169)
T. reesei	White pine	220	1.08	7.8	15.0	11.3	48	26.8	170)
T. reesei	Popular	200	0.5	.7.9	18.14	27.6	24	100	171)
B. In vivo									
C. thermocellum	Hardwood	220	1.0	9.0	4.14	7.0	12	86	169)
C. thermocellum	Hardwood	220	1.0	9.0	5.48	2.74	16	88.6	169)

[a] "Hardwood" is 90% birch, 10% maple;

[b] calculated for data from [170] using values given for unpretreated solids assuming all hemicellulose and 10% of lignin are solubilized;

[c] calculated based on free cellulase activity for continuous cultures. Activity for *C. thermocellum* is measured on Avicel, and activity of *T. reesei* cellulase is measured on filter paper. However, the activity of both *C. thermocellum* and *T. reesei* cellulases are about equal on Avicel and filter paper [151];

[d] reaction time: time of experiment for batch experiments; residence time for continuous experiments:

[e] all yields are based on the glucose content of the solids before reaction

gues. In vitro and in vivo hydrolysis data are presented in Table 10, and compared with data for the cellulase of *Trichoderma reesei*. In vitro hydrolysis yields at comparable enzyme loadings and reaction times are somewheat higher for the *C. thermocellum* cellulase system than for *T. reesei* cellulase acting on both pretreated mixed hardwood and pretreated white pine. The substrate concentrations are however lower for the results with *C. thermocellum*. The difference between the yields on pine and mixed hardwood attests to the difficulty of effectively pretreating softwood previously mentioned (Sect. 2.2.2). Poplar hydrolysis by *T. reesei* cellulase is included in Table 10 to illustrate how much more easily this substrate is hydrolyzed by this system than is mixed hardwood. Hydrolysis of poplar using *C. thermocellum* cellulase has not been studied. Yields from in vivo cellulose hydrolysis by *C. thermocellum* are slightly lower than in vitro yields using *C. thermocellum* cellulase, and are essentially equal to the yields obtained with *T. reesei* cellulase at comparable reaction times. The lower apparent enzyme loading, based on free cellulase, in the in vivo studies may be indicative of most of the enzyme being bound to cells or substrate, and/or a greater effectiveness of the enzyme in the presence of cells.

The data in Table 10 suggest that high extents of substrate utilization should be attainable with thermophilic systems and pretreated hardwood substrates. Demonstration of comparable performance at high substrate concentrations represents an important goal for future applied research in this area. Dilute-acid pretreatment of softwood substrates appears unlikely to be a successful pretreatment for this enzyme system, as with others. Testing *C. thermocellum* cellulase activity against substrates pretreated by means other than dilute acid hydrolysis would be informative.

4.2 Utilization of Pentose Sugars

The ability of thermophiles to consume a very broad range of carbohydrates, including pentose sugars and their polymers, is well established. This ability is also a key advantage of thermophiles for production of ethanol from lignocellulose. In particular, the simultaneous utilization of the hexose and pentose sugars present in biomass is desirable from a practical point of view. Thermophiles such as *Clostridium thermohydrosulfuricum*, *C. thermosaccharolyticum*, whose taxonomic status has been questioned [10], and *Thermoanaerobacter ethanlicus* utilize xylose and other pentose sugars at rates comparable to hexoses [8, 10]. Notably, all of these species are not cellulolytic, and the cellulolytic *C. thermocellum*, does not utilize pentoses [11]. Thus simultaneous utilization of pentose and hexose sugars as present in pretreated biomass by described species of thermophilic bacteria requires that both a cellulolytic and pentose-utilizing organism be present. Cocultures have been studied with *C. thermocellum* paired with *C. thermosaccharolyticum* [130, 168], *C. thermohydrolsulfuricum* [164, 168, 172], and *T. ethanolicus* [7, 172]. The role of the non-cellulolytic bacterium in these co-cultures, reviewed by Carreira and Ljungdahl [7], includes hexose as well as pentose utilization. This is suggested by marked effects of the presence of the non-cellulolytic organism in cultures grown on pentose-free substrates, and also the general tendency for end-product profiles in co-cultures to more nearly reflect that of the non-cellulolytic bacterium.

Simultaneous utilization of hexose and pentose sugars has been investigated to a limited degree in batch culture, and not at all in continuous culture. Slaff and Humphrey [173] report diauxic utilization of glucose in preference to xylose or cellobiose for *C. thermohydrosulfuricum*, with xylose and cellobiose consumed simultaneously in the absence of glucose. Carreira et al. [174] found that glucose and xylose are used simultaneously by *T. ethanolicus*, but that glucose is used in preference to starch. It may be noted that *C. thermocellum* uses cellobiose in preference to glucose [175]. Limited data are available on pentose and hexose utilization during growth on insoluble substrates. Avgerinos and Wang [130] reported 85—90% utilization of both glucose and xylose at the end of batch conversion of solvent-extracted corn stover, indicating that both carbohydrate species were eventually used to a large extent.

4.3 Ethanol Tolerance

The ethanol tolerance of thermophilic bacteria is an important applied characteristic which, along with ethanol yield, is usually seen as the greatest barrier to the commercial utilization of thermophiles for ethanol production [14,166,176]. Slapack et al. [8] and Lovitt et al. [10] have reviewed strain development to increase ethanol tolerance in thermophilic bacteria, and Slapack et al. [8] and Lovitt et al. [10] and Rogers [3] have reviewed mechanisms of ethanol tolerance and sensitivity.

The mechanisms of ethanol tolerance and intolerance in thermophiles may be different for different species. Increasing ethanol resistance with decreasing temperature has been reported for both *C. thermocellum* [177] and *C. thermohydrosulfuricum* [178]. This observation was used to support an explanation for ethanol inhibition in terms of membrane fluidity in the former study, but not in the latter. Slapack et al. [8] discuss membrane fluidity effects in detail. Herrero et al. [179] attribute ethanol intolerance to a blockage in glycolysis, possibly in response to ethanol-induced changes in the membrane [180,181]. However, Lovitt et al. [10] concluded that ethanol inhibition was not due to either of these effects but is due to regulatory phenomena. It may be noted that Lovitt et al.'s interpretation of ethanol sensitivity is based in part on greater sensitivity to ethanol than to other solvents (methanol and acetone). Though the tolerance of *C. thermocellum* to acetone and methanol has not to the authors knowledge been published, this organism does exhibit different solvent sensitivity, with sensitivity to ethanol greater than to either propanol or butanol [182].

Lovitt et al. [10] have recently proposed a mechanism for ethanol inhibition in *C. thermohydrolsulfuricum*, which also is consistent with the sensitivity of this organism to inhibition by hydrogen. In the presence of inhibitory concentrations of either ethanol or H_2, the NADH/NAD ratio increases by a factor of about 2.5 in wild-type cells, but not in an ethanol-resistant mutant. Addition of acetone provides an acceptor for electrons carried by NADH, and relieves inhibition. These effects have been attributed to inhibition of glycolysis, in particular glyceraldehyde dehydrogenase, due to over reduction of the cellular pyridine nucleotide pool [10].

Though the ethanol-tolerance of wild-type thermophilic bacteria is typically <1% ethanol, numerous workers have selected ethanol-tolerant strains [see [8]]. Mutants with quite high ethanol tolerance have been reported. Examples include a strain of *T. ethanolicus* able to tolerate 10% ethanol [7], *C. thermocellum* S7, which can tolerate

6% ethanol with <50% inhibition of growth (the basis is not clear) [14], and *C. thermohydrosulfuricum* 39EA [178], which grows well at 8% ethanol at 45 °C, and 5% ethanol at 60 °C.

There is a substantial difference between reports of the concentrations of ethanol tolerated by thermophiles and the concentrations of ethanol they produce. In a Masters thesis completed in 1982, Kim [183] observed production of 60 g L^{-1} ethanol by *C. thermosaccharolyticum* growing on xylose in fed-batch culture. However, more than half this concentration was produced after the cessation of exponential growth, and ethanol production occurred over a period of nearly 200 h. Writing in 1983, Carreira et al. [174] state that the highest concentration of ethanol reported to be produced by thermophilic anaerobes from xylose, glucose or starch is ~3% v/v ethanol. *C. thermocellum* is generally considered to be less ethanol tolerant than the frequently-studied non-cellulolytic thermophiles [7,10]. Ethanol concentrations of 1.5–1.7% have consistently been produced by batch co-cultures of *C. thermocellum* and *C. thermosaccharolyticum* (personal communication, A. Demain, MIT). The recent and comprehensive reviews of Slapack et al. [8] and Lovitt et al. [10] mention no values higher than 3–4% ethanol produced by thermophiles in batch culture, and it is not stated what fraction of this is formed after growth ceases. These values may be compared to the maximum endogenously-produced ethanol concentration allowing cell growth in yeast, which is about 9 wt.% [85]. In reviewing strain development work at MIT in 1986, Mistry [184] states that available strains of either *C. thermocellum* or *C. thermosaccharolyticum* are unable to grow rapidly in the presence of ethanol concentrations >20 g L^{-1}, and describes a difference of 2-fold in the inhibition of growth by exogenous and endogenous ethanol. Mistry also reports rather severe problems with revertance for an ethanol tolerant mutant of *C. thermocellum* strain S7. The highest ethanol concentration known to the author produced by thermophiles in continuous culture is 2 to 2.4 wt.%, reported by Kim [183] for *C. thermosaccharolyticum*. In work to be published with the same organism, the author and colleagues have observed values of 1.5 to 1.7 wt.% ethanol in continuous cultures. Continuous production of ~1% ethanol has been reported for *C. thermosaccharolyticum* [184], and *C. thermohydrosulfuricum* [185].

Within the thermophilic ethanol tolerance literature, reported ethanol tolerance is generally highest for exogenously added ethanol, intermediate for endogenously-produced ethanol formed in batch culture, often with extensive ethanol formation after cessation of growth, and lowest for ethanol produced by exponentially-growing cells. Distinguishing between these measures of ethanol tolerance is important because typical values for the various measures differ by a factor of at least 3, and perhaps 5. Realistically interpreting ethanol tolerance and production data for thermophilic bacteria is made more difficult by the fact that experiments in which high ethanol concentrations are obtained are often not presented in detail but are referred to as unpublished results or personal communication. Instability of ethanol-tolerant strains further complicates evaluation [184].

The concentration of endogenously-produced ethanol is a more relevant measure of ethanol production capability from a practical perspective than is exogenously-tolerated ethanol. In some cases relative tolerance to exogenous ethanol may indicate a potential not yet achieved in studies on produced ethanol. However, it is well established that yeast is more tolerant to exogenously added ethanol than to endogenous ethanol [186]; the same appears to be true of thermophiles [184]. The ethanol concentration which can be achieved in free-cell continuous culture is that produced and tolerated by exponentially-growing cells. Higher ethanol concentrations have been observed and are likely to be achieved in batch culture. However, batch culture systems for ethanol production using thermophilic bacteria can be expected to have lower productivity than continuous systems, and are not easily coupled with continuous ethanol removal (see Sect. 3.1). Cell immobilization and cell recycle systems may allow relatively high ethanol concentrations to be produced while enjoying some of the advantages of continuous systems. These systems are generally less easily implimented when insoluble substrates are employed. Differential residence times for liquid and insoluble substrates in a continuous system are relatively easy to achieve, and may provide a means of retaining cells attached to the substrate.

4.4 End-Product Metabolism and Ethanol Yields

Metabolic control and manipulation of end-product yields, and the selection of mutants with high ethanol yields are reviewed by Slapack et al. [8], Rogers [3], and Lovitt et al. [10]. The branched pathway leading to ethanol, acetic acid, and lactic acid is shown in Fig. 6. Considerations pertaining to end-product metabolism are similar for hexose and pentoses utilization because end-products are formed via pyruvate from both classes of substrates [3]. The carbon intermediates in this pathway are generally the same for different species of thermophilic anaerobes, however the electron carriers and also demonstrated reversibility for particular reactions varies between species and even strains [187, 188]. It is theoretically possible for ethanol, acetic acid or lactic acid to be the sole soluble product in that the pathways leading to all of these products result in balanced production and consumption of electron carriers and net generation of ATP. For production of acetate only, the NADH produced by glycolysis could be reoxidized with the concomitant reduction of ferredoxin and subsequent production of hydrogen gas.

Two classes of regulatory mechanisms may be involved in thermophilic endproduct metabolism: those which regulate end-product synthesis in response to the flux of intermediates in the pathways leading to particular end-products, and those which respond to the concentration of the end-products. These mechanisms are not mutually exclusive, and may both have a role. The relative importance of these mechanisms is important not only scientifically, but also practically. Concentration-dependent mechanisms should allow higher ethanol yields by manipulation of end-product concentrations in the bioreactor environment, for example by ethanol removal, whereas flux-dependent mechanisms should not allow this.

In a classic paper, Thauer et al. [189] review end-product control of anaerobic metabolism. They explain end-product distributions in terms of regulation of metabolic efficiency by controlling the flux of intermediates in pathways leading to products with different ATP yields. Mistry [190] has recently employed intrinsically-based mechanisms in explaining the end-product yields of *C. thermosaccharolyticum*.

Control of the rate of reactions in enzyme-catalyzed metabolic pathways is different for reversible reactions than for essentially irreversible reactions [191]. The rates of

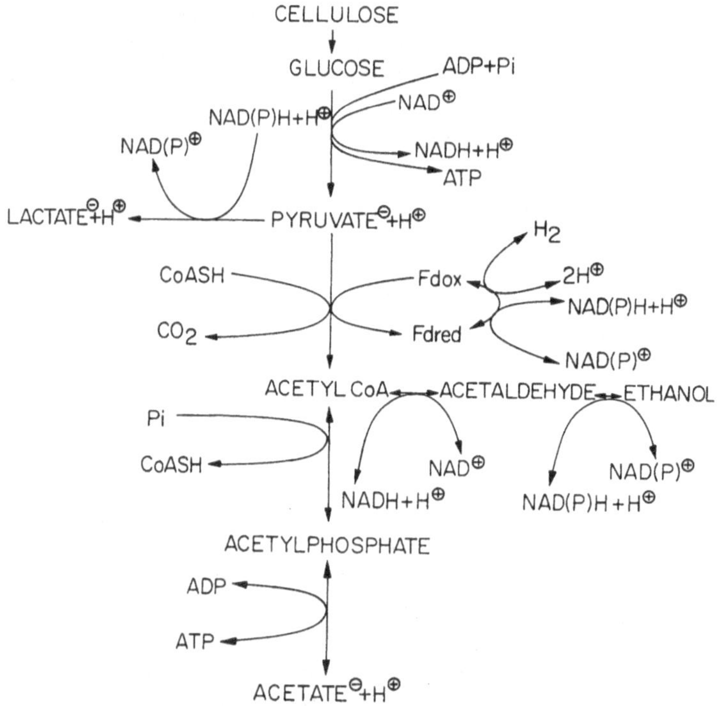

Fig. 6. Typical end-product metabolism of thermophilic ethanol-producing bateria data from [7, 187, 188]. Reactions shown as reversible have been demonstrated to be reversible in at least one thermophilic species

reversible reactions are controlled by substrate and product concentrations. The rates for essentially irreversible reactions are controlled by enzyme activity, and are frequently subject to allosteric control. As shown in Fig. 6, the majority of the reactions involved with end-product metabolism in thermophiles have been shown to be reversible in at least some organisms. Many, and for some organisms all, of the reactions for interconversion of the electron carriers NAD, NADP, and ferredoxin have been shown to be reversible. Thus the concentration of end-products can be expected to exert an effect on the pattern of end-products formed. Several examples of such effects are discussed below.

According to Slapack et al. [8], the most important factors in regulating end-product ratios in thermophilic anaerobes are alcohol dehydrogenase, ferredoxin: NAD(P) oxidoreductases, and FDP-activated lactate dehydrogenase. The role of hydrogen is discussed by Rogers [3]. Generally, increasing hydrogen partial pressure increases ethanol yield, though the sensitivity to hydrogen depends on the activity of ferrodoxin: NAD(P) oxidoreductase. Dramatic increases in ethanol yield due to increased hydrogen partial pressure have been reported for *C. thermohydrosulfuricum* [195], *C. thermosaccharolyticum* [184], and *C. thermocellum* [192]. All of these reports describe supersaturation of the growth medium with hydrogen, and changes in the ethanol yield by variables which would be expected to effect the degree of hydrogen supersaturation, e.g. gas sparging and stirring rate. Lovitt et al. [10] discuss the differences in response to hydrogen amoung thermophiles and possible mechanistic explanations for these differences.

Mass-action effects involving soluble end-products have also been implicated in determining end-product ratios. Herrero-Molina [182] found that addition of each of the fermentation products of *C. thermocellum* (ethanol, acetate, and lactate) results in decreased synthesis of the added product and increased synthesis of the other products. Altered end-product ratios due to selective addition or removal of a particular product has also been reported in other systems involving thermophilic ethanol producing bacteria [193,194].

As discussed in Sect. 3.1, the added costs associated with operating at low ethanol concentration may be limited to ethanol separation per se by employing continuous ethanol removal. In addition, ethanol removal may be expected to increase ethanol yield if concentration-dependent control mechanisms are operative. Only preliminary lab-scale studies of continuous ethanol removal from thermophilic systems have been undertaken, some of them without proper regard for energy requirements. In addition to careful consideration of energy requirements, it is also important to evaluate the physiological impact of proposed ethanol removal techniques. Though ethanol removal may have the potential to increase ethanol yields, the opposite effect has been observed. Sundquist et al. [185] found that ethanol removal via a reduced pressure flash vessel decreased ethanol yields due to reduced hydrogen partial pressure. In reduced-pressure ethanol removal systems, minimization of the time spent in the ethanol removal apparatus is likely to be critical.

Increasing iron concentration has been found to promote ethanol production in *C. thermosaccharolyticum* [195], a result explained in terms of limitation of ferredoxin synthesis and thus pyruvate dehydrogenase activity. However, Ljungdahl et al. [196] report using iron limitation to obtain a strain of *T. ethanolicus* with high ethanol yields. Growth rate has also been implicated in controlling ethanol yield. Zeikus et al. [128] and Ben-Bassat et al. [194] report high ethanol yields at low growth rates with both *Thermoanaerobium brockii* and *C. thermohydrosulfuricum*. Avgerinos and Wang [130] were able to obtain high ethanol yields only at high growth rates using *C. thermosaccharolyticum* and *C. thermocellum* in both mono- and co-culture.

Thermophilic bacteria, including both wild-type and mutant strains, have been reported with high ethanol yields. For example, an ethanol yield of 1.95 moles $mole^{-1}$ glucose has been reported for *T. ethanolicus* JW200 [174], and a yield of 1.9 moles $mole^{-1}$ glucose has been reported for *C. thermohydrosulfuricum* 39E [197]. (These yields are both higher than would be expected based on incorporation of 10% of the substrate into cell material, which may indicate uncoupled growth and substrate consumption). However, substantially lower yields have also been obtained using both of these species, including the same strains [172,173,198]. Unfortunately, ethanol yields appear to be rather sensitive to growth conditions. In particular, several studies have found ethanol yields to be higher on laboratory substrates than on heterogeneous substrates derived from native biomass [130,172]. Personal communication with several investigators suggest that high-yielding strains do best in the hands of the investigators that isolated them. This observation also applies to ethanol tolerant mutants. Mutants with ratios of ethanol to acidic end-products of >5:1 have been obtained for both *C. thermocellum* (strain S-7, [14]) and *C. thermosaccharolyticum* (strains HG6 and HG8, [130,199,200]). However, neither of these strains perform so well in the hands of the author. Dr. Demain's group at MIT found the reduced acetate production reported for a mutant strain of *C. thermosaccharolyticum* [201] to be an unstable characteristic.

From a mechanistic point of view, it is very clear that concentration-dependent metabolic control mechanisms are operative in thermophilic ethanol-producing bacteria. The situation is not so clear with respect to flux-dependent mechanisms. At the very least, the "set point" in terms of efficiency of energetic coupling appears to be very flexible in light of the drastically different product yields produced by different thermophilic species and strains. A definitive work on ethanol removal from thermophilic cultures might be expected to shed further light on this question, but has yet to appear in the literature.

From a practical point of view, the goal of obtaining stable thermophilic strains producing ethanol at high yields and relatively high concentrations from practical substrates has not been entirely achieved to the author's knowledge. Development of cellulolytic strains with these properties would appear to be a particular priority for research.

5 Concluding Remarks

Notwithstanding short term economic factors and trends in the attention of scientists and policy makers, the prospect of decreasing and ultimately vanishing supplies of oil is a real problem today for the same reasons it was a real problem five years ago. Moreover this problem is not one confined to the distant future. In countries such as the United States, where demand for oil is particularly high relative to reserves, large increases in the real oil price have been forecast within the next 15 years [22]. Strategies aimed at maintaining oil reserves, boosting domestic production, and maintaining good relationships with oil-rich countries can buffer the effects of oil exhaustion in oil-poor countries. However, developing alternatives to oil is the only real solution.

Biological production of ethanol is worthy of attention today as a route to partially replacing oil for the same reasons this process received considerable attention five years ago. Ethanol is a versatile chemical feedstock, and an excellent fuel which can displace gasoline on a better than 1:1 basis in energetic terms. Furthermore biological systems are well-known for their desirable catalytic features including high selectivity and reaction under mild conditions [202]. However, in order to make a significant contribution to replacing oil, ethanol production systems must convert a plentiful feedstock in a manner which is acceptable from economic, environmental, and energetic points of view. From this perspective, production of ethanol from starch- or sugar- rich substrates, as is presently practiced, has limited potential. Efforts to develop processes for ethanol production from substrates other than sugar- and starch-rich agricultural crops have not been successful to date, though a decade of intensive research has provided many insights of both fundamental and applied value.

The quantity of fermentable carbohydrate available for ethanol production from lignocellulosic substrates in the United States appears to be sufficiently large that conversion of this material into ethanol to replace liquid transportation fuels is a realistic possibility in this country. Biomass supplies are on average larger relative to oil consumption in countries other than the U.S. Compared to corn, the potential ethanol production from lignocellulosic substrates in the U.S. is greater by a factor on the order of 20. At ethanol production levels above that which can support the corn by-product market as an animal feed, lignocellulosic substrates potentially have superior features to corn from energetic and environmental viewpoints, and essentially do not compete with food production. In addition to these advantages, lignocellulosic substrates can be distinguished from starch- and sugar-rich substrates by their relatively high content of insoluble polymers containing β-linked glucose (cellulose) and pentoses (found in hemicellulose). The difficulty in economically converting these components has been primarily responsible for the incomplete success to date in efforts to develop practical processes for ethanol production from lignocellulose.

Very little research on thermophilic ethanol production has been reported under conditions which approach being practical. However, the value of distinguishing features of thermophilic bacteria for ethanol production can be assessed in economic terms in order that the potential of these organisms can be more clearly defined, and priority research areas identified. Using this approach, this study concludes that thermophilic bacteria offer the potential for a large, e.g. two-fold, reduction in ethanol production costs compared to processes based on available technology using enzymatic or acid hydrolysis and yeast. These yeast-based processes may of course be improved [70,78], however they are likely to have several more process steps, and other things being equal higher cost, in the absence of breakthroughs in substrate utilization capability. The primary potential advantages of thermophilic bacteria compared to yeast for conversion of lignocellulosic substrates arise from their ability to utilize cellulose and pentoses, and not from process advantages conferred by thermophily. Thus thermophilic bacteria are particularly suited to lignocellulosic substrates rather than starch- and sugar-rich substrates which can be made convertable by yeasts at relatively low cost.

In evaluating thermophilic bacteria for ethanol production, it has frequently been assumed that ethanol tolerance comparable to yeast is required. Research aimed at strain development to achieve such ethanol tolerance has been unsuccessful to date. The requirement for high ethanol tolerance is likely to apply if separation technologies which are designed for the ethanol concentrations tolerated by yeasts are employed. One general conclusion of the present study is that product recovery technologies which have higher capital costs but low energy requirements for separation of dilute ethanol solutions can have small cost penalties in terms of overall process economics. In particular, these costs are small relative to the reduction of total costs resulting from the key advantages of thermophiles for ethanol production from lignocellulose.

Engineering-based approaches appear able to lower the threshold ethanol concentration for a practical process into the range of concentrations produced by actively growing thermophilic cultures. It is very likely that approaches to dealing with the problem of low ethanol yields will be based to a much larger extent on biochemical understanding of the metabolism of thermophilic bacteria. Both strain development, using a variety of screening and selective techniques, and manipulation of environmental conditions have been somewhat successful in obtaining high ethanol yields under some conditions. However, stability of high ethanol yielding strains has been a problem with respect to both repeatability of results and sensitivity to growth conditions. Moreover, few strains with high ethanol yield have been tested under repeated sub-culturing or continuous culture.

In general, process-oriented studies of thermophilic ethanol production are scarce in the literature, and much needed if development of this technology is to proceed. Particularly lacking are studies in continuous culture, studies at high concentrations of economically interesting feedstocks containing a combination of substrates and also potentially inhibitory compounds, and ethanol removal studies.

The material and analysis presented in this paper lead the author to conclude that there is a large gap between the potential of thermophilic bacteria for ethanol production, even at relatively low ethanol concentrations, and that which has been experimentally demonstrated. Process studies as described above and research to obtain stable thermophilic cultures with high ethanol yields under realistic conditions would appear to be particularly important topics for study in order to close this gap.

6 Acknowlegements

I am grateful to the many people who contributed to this review by generously making available their reference collections, original papers, and ideas. These include Drs. H. W. Blanch, A. O. Converse, C. L. Cooney, R. Datta, A. L. Demain, H. E. Grethlein, C. High, L. G. Ljungdahl, L. Mednick, T. Peterson, K. Venkatasubramanian, R. H. Wolkin, and J. G. Zeikus, and also R. Chamberlin, C. Leach, K. Levinson, and G. Maffia. I thank C. Johnson and D. Call for donating their time and skill in the preparation of figures. Also I thank Drs. Converse, Wolkin, S. C. Lynd, and also J. Byrnes, for critically reading and proof-reading the manuscript. Finally I thank Drs. A. E. Balber, H. E. Grethlein, N. J. Poole, and J. G. Zeikus for inspiration and encouragement.

7 References

1. Cowling EB, Kirk TK (1976) Biotechnol. Bioeng. Symp. 6: 95
2. Lamed R, Bayer EA (In press) The cellulosome of *Clostridium thermocellum*. In: Laskin AI (ed) Advances in applied microbiology. Academic, NY, vol 33
3. Rogers P (1986) Genetics and biochemistry of *Clostridium* relevant to development of fermentation processes. In: Laskin AI (ed) Advances in applied microbiology. Academic, New York, vol 31 p 1
4. Hartley BS, Payton MA (1983) Biochem. Soc. Symp. 48: 133
5. Sonnleitner B (1983) Biotechnology of thermophilic bacteria — growth, products and application. In: Fiechter A (ed) Advances in biochemical engineering/Biotechnology. Springer, Berlin Heidelberg New York, vol 28 p 69
6. Sonnleitner B, Fiechter A. (1983) Trends Biotechnol. 1(3): 74
7. Carreira, LH, Ljungdahl LG (1984) Production of ethanol from biomass using anaerobic thermophilic bacteria. In: Wise DL (ed) Liquid fuel developments (CRC series in biotechnology) CRC, Boca Raton FL
8. Slapack GE, Russel, I, Stewart GG (1987) Thermophilic microbes in ethanol production. CRC, Boca Raton FL
9. Wiegel J, Ljundahl LG (1986) CRC Crit. Rev. Biotechnol. 3(1): 39
10. Lovitt, RW, Kim, BH, Shen G-J, Zeikus JG (In press) Solvent production by microorganisms (to be published in CRC Critical Reviews)

11. Duong T-VC, Johnson EA, Demain AL (1983) Thermophilic anaerobic cellulolytic bacteria. In: Wiseman A (ed) Topics in enzyme and fermentation biotechnology. vol 7 p 156 Halsted Press, Hopwood N.Y.
12. Ljungdahl LG, Eriksson KE (1985) Ecology of microbial cellulose degradation. In: Marshall KC (ed) Advances in microbial physiology, vol 8 p 237 Academic Press, N.Y.
13. Wiegel J (1982) Experentia 82: 151
14. Wang DIC, Averginos GC, Biocic I, Fang SD, Fang HY (1983) Philos. Trans. R. Soc. London B300: 323
15. Grathwohl M (1982) World energy supply. de Gruyter, Berlin
16. Vergara W, Pimentel D (1978) A study of the energy potential of fuels from biomass in five countries. In: Energy from biomass and wastes. Symposium, 14–18 May, Washington DC. Available from The Institute, Chicago
17. Owsley DC, Bloomfield JJ (1985) Chemtech 15(2): 94
18. Exxon Corporation (1981) World energy outlook. Corporate Planning and Public Affairs Department, New York
19. National Petroleum News Fatbook Issues, 1984–1987
20. American Petroleum Institute (1988) Basic petroleum data book — Petroleum Industry Statistics, *8*(1), Washington
21. Mast RF, Dolton GL, Crovelli RA, Powers RB, Charpentier RR, Root DH, Attanasi ED (1988) Estimates of undiscovered recoverable oil and gas resources for the onshore and state offshore areas of the United States. In: USGS Program and Abstract on Mineral and Energy Resources, V.E. McKelvey Forum, Denver CO, Abstract in the U.S. Geolgical Survey Circular 1025
22. U.S. Department of Energy (1985), National energy policy plan projections. Office of Planning and Analysis; DOE/PE-0029/3
23. — Chemical marketing reporter. (1960 through 1986) Schnell Publishing, New York
24. Peterson T Wisconsin forest extension price reviews. Cooperative Extension Program, U.S.D.A., Madison, Wisconsin (April 1981 through November 1986)
25. Rudderman FK (1980) Pacific northwest production prices employment and trade. In: Northwest Forest Industries. Pacific Northwest Forest and Range Experimental Station, Portland OR
26. Palsson BO, Fathi-Afshar S, Rudd DF, Lightfoot EN (1981) Science 213: 513
27. Ng, TK, Busche RM, McDonald CC, Hardy RWF (1983) Science 219(4585): 733
28. Busche RM (1985) Biotech. Prog. 1(3): 165
29. Hall DO (1979) Fuel 57(6): 322
30. Ferchak JD, Pye EK (1981) Solar Energy 26: 9
31. Energy from biological processes, volume II — technical and environmental analyses. Office of Technology Assessment, Congress of the United States, Washington DC (1980)
32. Humphrey AE, Moreira A, Armiger W, Zabriske D (1977) Biotechnol. Bioeng. Symp. 7: 45
33. Goldstein IS (1981) Biomass availability and utility for chemicals. In: Goldstein, IS (ed) Organic chemicals from biomass. CRC Press, Boca Raton, FL, p 1
34. Jeffries TW (1983) Utilization of xylose by bacteria, yeasts, and fungi. In: Fiechter A (ed) Advances in biochemical engineering/Biotechnology. Springer, Berlin Heidelberg New York, vol 27 p 1
35. Young J, Griffin E, Russell J (1986) Biomass 10: 9
36. Levinson A (1982) Resource Man. Optim. 2(2): 99
37. Venkatasubramanian K, Kiem C (1985) Starch and energy: technology and economics of fuel alcohol production. In: van Beynum GMA, Roels JA (eds) Starch conversion technology. Marcel Dekker, New York, p 143
38. Lipinsky ES (1978) Science 199: 644
39. Katzen R et al. (1978) Grain motor fuel alcohol technical and economic assessment. Available from NTIS, Springfield VA; HCP/J6639-01
40. Bellamy WD (1975) Conversion of insoluble agricultural wastes to SCP by thermophilic microorganisms. In: Tannebaum SR, Wang DIC (eds) Single cell protein II. MIT Press, Carelton, MA p 263
41. Stephens HR, Heichel GH (1975) Biotechnol. Bioeng. Symp. 5: 27

42. Gong C-S, Chen LF, Flickinger MC, Tsoa GT (1981) Conversion of hemicellulose carbohydrates. In: Fiechter A (ed) Advances in biochemical engineering/Biotechnology. Springer, Berlin Heidelberg New York, vol 20 p 93
43. Morrison FB (1956) Feeds and feeding — a handbook for the student and stockman. Morrison, Ithaca NY
44. Tchobanaglous GH, Theisen H, Eliassen R (1977) Solid wastes: engineering principles and management issues. McGraw-Hill, New York
45. Herman MF, Othmer DF, Overberger CG, Seaborg T (eds) Kirk-Othmer encyclopedia of chemical technology, 3rd edn, Wiley, New York (1978)
46. Gaines LL, Karpuk M (1987) Fermentation of lignocellulosic feedstocks: product markets and values. In: Klass DL (ed) Energy from biomass and wastes X, Institute of Gas Technology, Chicago
47. Maiorella BL, Blanch HW, Wilke CR (1983) Proc Biochem. 18(4): 5
48. Waits ED, Elmore JL (1983) Environ. Int. 9: 325
49. Loehr RC, Sengupta M (1985) Environ. Sanit. Rev. 16: 1
50. Badger Engineers Inc. (1984) Economic feasibility study of an acid-based ethanol plant. SERI, Golden, CO; ZX-3-03096-2
51. Chem Systems Inc. (1984) Economic feasibility study of an enzymatic hydrolysis-based ethanol plant with prehydrolysis pretreatment. SERI, Golden, CO; XX-0-03097-2
52. Matsuda S, Kubota H (1984) Biomass 4: 161–182
53. National Research Council, carbon dioxide assessment committee (1983) Changing climate. National Academy Press, Washington, DC
54. Hileman B (1984) Environ. Sci. Technol. 18(2): 45A
55. Rinehart S (1988) Stations start selling high-oxygen fuels. Colorado Daily, 94(284): 1
56. Hacking AJ (1986) Economic aspects of biotechnology. Cambridge studies in biotechnology 3. Cambridge University Press, Cambridge
57. Murtagh JE (1986) Process Biochem. 21(2): 61
58. Keim CR (1983) Enzyme Microb. Technol. 5: 103
59. Esser K, Karsch T (1984) Process Biochem. 19(3): 116
60. Greek BF (1987) Chem. Eng. News 65(6): 9
61. Smith N, Corcoran TJ (1981) Wood production energetics: an analysis for fuel applications. In: Klass DL (ed) Biomass as a nonfossil fuel source. ACS, Washington, DC (Symposium series No. 144), p 433
62. Ferchak JD, Pye EK (1981) Solar Energy 26: 17
63. Datta R (1981) Process Biochem. 16(4): 16
64. Wilke CR, Maiorella B, Sciamanna A, Tangnu K, Wiley D, Wong H (1983) Enzymatic hydrolysis of cellulose — theory and applications. Noyes Data Corp, Park Ridge, NJ
65. Grethlein H (1984) Biotech. Adv. 2: 43
66. Dale BE (1985) Cellulose pretreatments: technology and techniques. In: Tsoa GT (ed) Annual reports on fermentation processes, vol 8 p 299
67. Grethlein H (1985) Bio/Technol. 3(2): 155
68. Weimer PJ, Weston WM (1985) Biotechnol. Bioeng. 27: 1540
69. Grethlein H, Converse AO (1985) Understanding how pretreatment increases the rate of enzymatic hydrolysis of wood. Presented at: 190th meeting of the ACS
70. Wright JD, Power AJ, Douglas LJ (1986) Biotechnol. Bioeng. Symp. 17: 285
71. Allen DC, Grethlein HE, Converse AO (1984) Solar Energy 33(2): 175
72. Weimer PJ, Chou Y-CT, Weston WM, Chase DB (1986) Biotechnol. Bioeng. Symp. 17: 5
73. Preprints from: Symposium on the pretreatment of lignocellulosic materials. 23–27 June 1986, Graz, Austria. Forest Research Institute, Rotura, New Zealand
74. Ladisch MR, Lin KW, Voloch M, Tsoa GT (1983) Enzyme Microb. Technol. 5: 82
75. Mardsen WL, Gray PP (1986) CRC Crit. Rev. Biotechnol. 3(3): 235
76. Grethlein HE Acid hydrolysis review. Presented at: Conference on anaerobic digestion and carbohydrate hydrolysis of wastes. 8–10 May 1984, Luxembourg, Commission of the European Communities
77. Ladisch MR, Tsoa GT (1986) Enzyme Microb. Technol. 8: 66

78. Wright JD, Power AJ (1987) Comparative technicial evaluation of acid hydrolysis processes for conversion of cellulose to ethanol. In: Klass, DL (ed) Energy from biomass and wastes X. Elsevier, Essex
79. Smith PC, Grethlein HE, Converse AO (1982) Solar Energy 28(1): 41
80. Kwarteng K (1983) Kinetics of acid hydrolysis of hardwood in a continuous plug flow reactor. Ph.D. Thesis, Thayer School of Engineering, Hanover, NH
81. Parker S, Calnon M, Feinberg D, Power A, Weiss L (1983) The value of furfural/ethanol coproduction from acid hydrolysis processes. SERI, Golden, CO; TR-231-2000
82. Esser K, Schmidt U (1982) Process Biochem. 17(3): 46
83. Faust U, Prave P, Schlingmann M (1983) Process Biochem. 18(3): 31
84. Guidoboni GE (1984) Enzyme Microb. Technol. 6: 194
85. Maiorella BL, Blanch HW, Wilke CR (1984) Biotechnol. Bioeng. 26: 1003
86. Kolot FB (1984) Process Biochem. 19(1): 7
87. Mulder MHV, Smolders CA (1986) Process Biochem. 21(2): 35
88. Hoffman H, Scheper T, Schugerl K, Schmidt W (1987) Chem. Eng. J. (Lausanne) 34: B13
89. Matsumura M, Markl H (1984) Appl. Microbiol. Biotechnol. 20: 371
90. Crabbe PG, Tse CW, Munro PA (1986) Biotechnol. Bioeng. 28: 939
91. Cysewski GR, Wilke CR (1977) Biotechnol. Bioeng. 19: 1125
92. Ghose TK, Roychoudhury PK, Ghose P (1984) Biotechnol. Bioeng. 26: 377
93. Rogers PL, Lee KJ, Skotnicki ML, Tribe DL (1982) Ethanol production by *Zymomonas mobilis* In: Fiechter A (ed) Advances in biochemical engineering/Biotechnology. Springer, Berlin Heidelberg New York, vol 23 p 37
94. Montencourt BS (1985) *Zymomonas*, a unique genus of bacteria. In: Demain AL, Solomon N (eds) Biology of industrial microorganisms. Cummings, Menlo Park, p 261
95. Karsch T, Stahl U, Esser K (1983) Eur. J. Appl. Microbiol. Biotechnol. 18: 387
96. Magee, RJ, Kosaric N (1985) Bioconversion of hemicellulosics. In: Fiechter A (ed) Advances in biochemical engineering/Biotechnology. Springer, Berlin Heidelberg, New York, vol 32 p 61
97. Hartline FF (1979) Science 206: 41
98. Parkinson G (1981) Chem. Eng. 88(11): 29
99. Choudhury JP, Ghose P, Guha PK (1985) Biotechnol. Bioeng. 27: 1081
100. Essien D, Pyle DL (1983) Process Biochem. 18(4): 31
101. Garg DR, Ausikaitis JP (1983) Chem. Eng. Prog. 79(4): 60
102. Katzen R, Ackley WR, Moon, GD, Messick JR, Brush BF, Kaupisch KF (1981) Low energy distillation systems. In: Klass DL, Emert GH (eds) Fuels and chemicals from biomass. Ann Arbor Science, Ann Arbor
103. Busche RM (1984) Biotechnol. Bioeng. Symp. No. 13: 597
104. Barba D, Brandani V, Di Giacomo G (1985) Chem. Eng. Sci. 50(12): 2287
105. Schmitt D, Vogelpohl A (1983) Sep. Sci. Technol. 18(6): 547
106. Lee F-M, Pahl RH (1985) Ind. Eng. Chem. Process Des. Dev. 24: 168
107. Lynd LR, Grethlein HE (1984) Chem. Eng. Prog. 81: 59
108. Grethlein HE, Lynd LH (1986) U.S. Patent No. 4, 626, 321
109. Lynd LR, Grethlein HE (1986) AIChE J. 32(8): 1347
110. Martin SR (1982) Chem. Eng. NY 377: 50–53
111. Lyons TP (1983) Proc. Biochem. 18(2): 18
112. Kampen WH (1980) Hydrocarbon Process. 59(2): 72
113. Parker HW (1982) Mech. Eng. 104(5): 54
114. Parisi F (1983) Energy balances for ethanol as a fuel. In: Fiechter A (ed) Advances in biochemical engineering/Biotechnology. Springer, Berlin Heidelberg New York, vol 28 p 41
115. Pimentel LS (1980) Biotechnol. Bioeng. 22: 1989
116. Rothman H, Greenshields R, Calle FR (1983) The alcohol economy: fuel ethanol and the Brazilian experience. Francis Pinter, London
117. Sama DA (1981) Hydrocarbon Process. 60(7): 89
118. Yorifuji T (1981) Energy Dev. Jpn. 3: 195
119. Johnson MA (1983) Energy 8(3): 225
120. Krochta JM (1979) Energy analysis for ethanol from biomass. Second international conference on energy use management. Pergamon, New York, p 1956

121. Cooney CL, Mistry FR (1982) Analysis of direct fermentation of lignocellulose to ethanol. Presented at: 184th meeting of the ACS
122. Huibers DTA, Jones MW (1980) Can. J. Chem. Eng. 58: 718
123. Graff GM (1982) Chem. Eng. NY. 89(26): 25
124. Janshekar H, Fiechter A (1983) Lignin: biosynthesis, application, and biodegradation. In: Fiechter A (ed) Advances in biochemical engineering/Biotechnology. Springer, Berlin Heidelberg New York, vol 27 p 119
125. Clements LD, Beck SR, Heintz C (1983) Chem. Eng. Prog. 79(11): 59
126. Shah RB, Clausen EC, Gaddy JL (1984) Chem. Eng. Prog. 80(1): 76
127. Greek BF (1984) Chem. Eng. News 62(11): 17
128. Zeikus JG, Ben-Bassat ANgTK, Lamed R (1981) Thermophilic ethanol fermentations. In: Hollaender A (ed) Trends in the biology of fermentations for chemicals and fuels. Plenum, New York, p 441
129. Sonnleitner B, Cometta S, Fiechter A (1982) Biotechnol. Bioeng. 24: 2597
130. Avgerinos GC, Wang DIC (1983) Biotechnol. Bioeng. 25: 67
131. Sonnleitner B, Fiechter A, Giovanni F (1984) Appl. Microbiol. Biotechnol. 19: 326
132. Leschine SB, Canale-Parola E (1983) Appl. Environ. Microbiol. 46(3): 728
133. Sleat R, Mah RA, Robinson R (1984) Appl. Environ. Microbiol. 48(1): 88
134. Lawford GR, Lavers BH, Good D, Charley R, Fein J, Lawford HG (1982) *Zymomonas* ethanol fermentations: biochemistry and bioengineering. Presented at: International symposium on ethanol from biomass, 13–15 Oct 1982, Winnipeg, p 482
135. Lacis L, Lawford HG (1985) J. Bacteriol. 163(3): 1275
136. Fardeau M-L, Plasse F, Belaich J-P (1980) European J. Appl. Microbiol. 10: 133
137. Wiegel J (1982) Experientia 38: 151
138. Barras F, Boyer MH, Chambost JP, Chippaux M (1984) Mol. Gen. Genet. 197(3): 513
139. Gilkes NG, Langsford ML, Kilburn DG, Miller RC, Warren RAJ (1984) J. Biol. Chem. 259(16): 10455
140. Kotoujansky A, Diolez A, Boccara M, Bertheau Y, Andro T, Coleno A (1985) EMBO J. 4(3): 781
141. Skipper N, Sutherland M, Davies RW, Kilburn D, Miller RC, Warren A, Wong R (1985) Science 230(4728): 958
142. Shalita FP, Yablonsky MD, Dooley MM, Bucholz, SE, Kahrs SK, Murphy-Holland K, Eveleigh DE (1987) Genetic engineering of bacteria for alcohol fuel production. In: Klass DL (ed) Energy from biomass and wastes X, Elsevier, Essex, p 907
143. Imanaka T (1986) Application of recombinant DNA technology to the production of useful biomaterials. In: Fiechter A (ed) Advances in biochemical engineering/Biotechnology. Springer, Berlin Heidelberg New York, vol 33 p 1
144. Guthrie KM (1969) Chem. Eng. NY. 76(6): 114
145. Peters MS, Timmerhaus KD (1969) Plant design and economics for chemical engineers, 2nd edn, McGraw-Hill, New York
146. Lynd LR (1987) Production of ethanol from lignocellulosic materials using thermophilic bacteria. DE thesis, Thayer School of Engineering, Hanover, NH
147. Levy PF, Sanderson JE, Ashare E, Wise, DL, Molyneaux MS (1980) Liquid fuels production from biomass. Report for DOE/SERI contract no. AC02-77ET20050; DOE/ET/20050-T4
148. Bayer EA, Kenig R, Lamed RL (1983) J. Bacteriol. 156(3): 818
149. Ljungdahl LG, Pettersson B, Ericksson KE, Wiegel J. (1983) Curr. Microbiol. 9: 195
150. Ng TK, Zeikus JG (1981) Appl. Environ. Microbiol. 42(2): 231
151. Johnson EA, Sakajoh M, Halliwell G, Madia A, Demain AL (1982) Appl. Environ. Microbiol. 43(5): 1125
152. Spinnler HE, Lavigne B, Blachere H (1986) Appl. Microbiol. Biotechnol. 23: 434
153. Johnson EA, Reese, ET, Demain AL (1982) J. Appl. Biochem. 4: 64
154. Johnson EA, Bouchot F, Demain AL (1985) J. Gen. Microbiol. 131: 2303
155. Wu D, Demain AL (1986) Abstracts of the annueal meeting of the ASM, p 73
156. Wu D, Demain AL (1985) Abstracts of the annual meeting of the ASM, p 248
157. Hon-Nami K, Coughlan MP, Hon-Nami H, Ljungdahl LG (1986) Arch Microbiol. 145: 13
158. Afeyan N (1987) A mechanistic study of the *Clostridium thermocellum* cellulase system. PhD Thesis, MIT

159. Lamed RL, Setter E, Bayer EA (1983) J. Bacteriol. 156(2): 828
160. Lamed RL, Kenig R, Setter EA (1985) Enzyme Microb. Technol. 7: 37
161. Coughlan MP, Hon-Nami K, Hon-Nami H, Ljungdahl LG, Paulin JJ, Rigsby WE (1985) Biochem. Biophys. Res. Comm. 130(2): 904
162. Bisaria VS, Ghose TK (1981) Enzyme Microb. Technol. 3: 90
163. Lynd LR, Grethlein HE (1987) Biotechnol. Bioeng. 29: 92
164. Ng TK, Ben-Bassat A, Zeikus JG (1981) Appl. Environ. Microbiol. 41: 1337
165. Saddler JN, Chan MK-H (1982) Eur. J. Appl. Microbiol. Biotechnol. 16: 99
166. Kundu S, Ghose TK, Mukhopadhyay SN (1983) Biotechnol. Bioeng. 25: 1109
167. Khan AW, Asther M, Giuliano C (1984) J. Ferment. Technol. 62(4): 335
168. Saddler JN, Chan MK-H (1984) Can. J. Microbiol. 30: 2123
169. No. 141, and Wolkin, Lynd and Grethlein, manuscript in preparation.
170. Grethlein HE, Allen DC, Converse AO (1984) Biotechnol. Bioeng. 25: 1498
171. Knappert D, Grethlein HE, Converse A (1981) Biotechnol. Bioeng. Symp. 11: 66
172. Hon-Nami K, Coughlan MP, Hon-Nami H, Carriera LH, Ljungdahl LG (1985) Biotechnol. Bioeng. Symp. 15: 191
173. Slaff GF, Humphrey AE (1981) Diauxic growth of *C. thermohydrosulfuricum*. Presented at: 182nd meeting of the ACS
174. Carreira LH, Wiegel J, Ljungdahl LG (1983) Biotechnol. Bioeng. Symp. 13: 183
175. Ng TK, Zeikus JG (1982) J. Bacteriol. 150(3): 1391
176. Slater GJ, Wakelin WS (1985) Thermophilic ethanol fermentation: an engineering assessment. NTIS, Springfield, VA; PB85-169 142
177. Herrero AA, Gomez RF (1980) Appl. Environ. Microbiol. 40(3): 571
178. Lovitt RW, Longin R, Zeikus JG (1984) Appl. Environ. Microbiol. 48(1): 171
179. Herrero AA, Gomez RF, Roberts MF (1985) J. Biol. Chem. 260(12): 7442
180. Herrero AA, Gomez RF, Roberts MF (1982) Biochim. Biophys. Acta 693: 195 (1982)
181. Curatolo W, Kanodia S, Roberts MF (1983) Biochim. Biophys. Acta 734: 336
182. Herrero-Molina AA (1981) The physiology of *Clostridium thermocellum* in relation to its energy metabolism. PhD Thesis, MIT, Cambridge
183. Kim S (1982) Microbial production of ethanol by *Clostridium thermosaccharolyticum*. MS Thesis. MIT, Cambridge
184. Mistry FR (1986) Ethanol Production by *Clostridium thermosaccharolyticum* in a continuous cell recycle system. PhD Thesis, MIT, Cambridge
185. Sundquist JA, Blanch HW, Wilke CR (1986) Ethanol production with *Clostridium thermohydrosulfuricum*. Presented at: 192nd meeting of the ACS
186. van Uden N (1985) Ethanol toxicity and ethanol tolerance in yeasts. In: Tsao G (ed) Annual reports on fermentation processes, vol 8 p 11
187. Lamed R, Zeikus JG (1980) J. Bacteriol. 144(2): 569
188. Hyun HH, Shen G-J, Zeikus JG (1985) J. Bacteriol. 164(3): 1153
189. Thauer RK, Jungermann K, Dekker K (1977) Bacteriol. Rev. 41(1): 100
190. Mistry FR No. 178, and manuscript submitted for publication
191. Krebs H (1969) The role of equilibria in the regulation of metabolism. In: Horeker BL, Stadtman ER (eds) Current topics in cellular regulation, Academic, New York, vol 1 p 45
192. Su T, Lamed R, Lobos J, Brennan M, Smith J, Tabor D, Brooks R (1981) Bioconversion of plant biomass to ethanol. Final report for DOE subcontract no. XR-9-8271-1. SERI, Golden, CO
193. Weimer PJ, Zeikus JG (1977) Appl. Environ. Microbiol. 33(2): 289
194. Ben-Bassat A, Lamed R, Zeikus JG (1981) J. Bacteriol. 146: 192
195. Mistry F, Cooney CL (1985) Ethanol production by *Clostridium thermosaccharolyticum* in a continuous culture cell-recycle system. Presented at: 190th meeting of the ACS
196. Ljungdahl LG, Bryant F, Carriera L, Saiki T, Wiegel J (1981) Some aspects of thermophilic and extreme thermophilic anaerobic microorganisms. In: Hollaender A (ed) Trends in the biology of fermentation for chemicals and fuels. Plenum, New York p 397
197. Zeikus JG, Ben-Bassat A, Hegge P (1980) J. Bacteriol. 143: 432
198. Ward PJ, Matharasan R (1986) The effect of controlled redox potential on the growth and energetics of *Thermoanaerobacter ethanolicus*. Presented at: 192nd meeting of the ACS
199. Wang DIC, Dalal R (1986) U.S. Patent no. 4,568,644

200. Avgerinos GC (1982) Direct conversion of cellulosic biomass to ethanol by mixed culture fermentation of *Clostridium thermocellum* and *Clostridium thermosaccharolyticum*. Ph.D. Thesis, MIT, Cambridge
201. Rothstein DM (1986) J. Bacteriol. 165(1): 319
202. Bailey JE, Ollis DF (1977) Biochemical Engineering Fundamentals. McGraw-Hill, New York

Advances in Lignocellulosics Hydrolysis and in the Utilization of the Hydrolyzates

Federico Parisi
Istituto di Scienza e Tecnologia dell'Ingegneria Chimica dell'Università, via all'Opera Pia 15, I 16100 Genova, Italy

1 Introduction ... 54
2 Nature and Structure of Lignocellulosics ... 54
3 Acid Hydrolysis ... 55
3.1 History ... 55
3.2 Theory of Acid Hydrolysis ... 55
3.3 The State of the Art ... 58
4 Enzymatic Hydrolysis ... 63
4.1 The *Trichoderma* Route ... 63
4.2 Other Cellulolytic Microorganisms ... 65
4.3 The Pretreatment ... 66
4.4 Enzyme Production and Hydrolysis Processes ... 69
5 The Utilization of Hydrolyzates and By-products ... 70
5.1 Ethanol Production ... 70
5.1.1 Ethanol from Pentoses ... 71
5.2 Furfural Production and Utilization ... 72
5.3 Other Uses of the C_5 Hydrolyzates ... 73
5.4 The Utilization of Lignin ... 73
5.5 Other Utilizations of Hydrolyzates ... 74
6 Ethanol Production Economics ... 75
7 Conclusions ... 80
8 References ... 80

The debate if acid or enzymatic hydrolysis of lignocellulosics will prevail over the other in the near future is still open. Different types of acid hydrolysis (use of concentrated acids, or diluted acids, and in this case use of extreme temperatures, or attempts to realize semi-continuous or continuous processes) are described. Advantages and inconveniences are described for each case.

However, only a limited margin for improvement is left to acid hydrolysis, compared to the enzymatic one. New microorganisms, new strains, and genetic engineering are actually improving classic enzymatic processes. Simultaneous hydrolysis and utilization of produced sugars will substantially modify current perplexities. A survey of the present trends is given.

In any case, the utilization of lignocellulosics hydrolysis will be of commercial interest only if hemicellulose hydrolyzates and lignin find profitable employment. Considerable effort is being made in this direction by both the scientific and the technical community.

1 Introduction

The literature on the use of lignocellulosic biomass, especially on the hydrolysis and biotransformation of cellulose and hemicellulose to ethanol and the utilization of the products and byproducts, is extremely abundant. This wealth is justified by the enormous quantity of this biomass available. The available quantity of lignocellulosic biomass in 1972 was estimated to be one hundred billion (10^{11}) tons per year [1]. The current value may very well be higher.

Many articles from this series are devoted to the theme of exploiting lignocellulosic biomass [1-20]. The specific topics treated are hydrolysis and its products utilization, with emphasis on the large-scale production of ethanol [21]. A complete survey of the most recent literature is beyond the scope of this article: here we will summarize the present trends and the progress being made toward industrial-scale applications.

2 Nature and Structure of Lignocellulosics

Lignocellulosic nature and structure have been treated in numerous publications, including issues in this same series [6,9,10,19]. Only the fundamentals are summarized here.

Lignocellulosic materials are composed of cellulose, hemicellulose, and lignin. The respective quantities in the different species are indicated in Table 1. Cellulose,

Table 1. Average composition of lignocellulosics

Species	Cellulose	Hemicelluloses	Lignin
Conifers	40–50%	20–30%	25–35%
Deciduous trees	40–50%	30–40%	15–20%
Cane bagasse	40%	30%	20%
Corn cobs	45%	35%	15%
Corn stalks	35%	25%	35%
Wheat straw	30%	50%	15%

a polymer of β-glucose, has a high degree of polymerization (200 to 2000 kDa) and crystallinity. Hemicelluloses are polymers of pentoses (xylose, mainly, and arabinose and ribose), hexoses (as glucose, mannose and galactose), and uronic acids. Lignin is made up of phenylpropane units, methoxylated and linked in various ways. It acts as a binder among the fibers in the lignocellulosic materials; thus, wood can be considered a natural example of composite material.

The high crystallinity of the cellulosic material is the first, fundamental obstacle to its hydrolysis, whatever the method used. Also, the lignin and hemicellulose shield the cellulose from enzymatic attack. In order to make an economically competitive process, it is necessary to convert each of these three major components into saleable products.

3 Acid Hydrolysis

3.1 History

Braconnot, in 1812, first attempted to hydrolyze cellulose using concentrated sulfuric acid. In 1913, Willstätter hydrolyzed cellulose with fuming hydrochloric acid and hydrolysis with diluted sulfuric acid was first performed by Simonsen in 1888.

Only at the beginning of the twentieth century, however, was industrial-scale hydrolysis considered for the production of sugar solutions. A plant using the diluted sulfuric acid process of Ewen and Tomlison [22,23] was built in the USA in 1910. Several generations of process improvement followed, culminating in the "Scholler" [24–27] and "Madison" [28] percolation processes. A Scholler-type plant used sulfuric acid at 0.6% at a maximum temperature of 184 °C and produced a solution at about 3% of glucose, with a yield of about 50% on cellulose. The Madison process had cellulose conversions of 50% and sugar concentrations of 4 to 5%, under the same conditions of temperature and concentration of the acid [29].

A commercial scale concentrated acid process using hydrochloric acid, was constructed by Bergius in 1925 and described later [30]. Research has also been carried out with concentrated sulfuric acid in Japan, and in the USA, and with hydrofluoric acid in Germany [31,32].

The primary advantages of acid hydrolysis are its rapid rate and simplicity. Concentrated acid hydrolysis has the advantage of high yields (virtually 100%), but suffers from high acid consumption and/or recovery costs. Dilute acid hydrolysis is extremely simple, but has low yields and produces large amounts of degradation products which inhibit microbial activity.

3.2 Theory of Acid Hydrolysis

The proposed hydrolysis mechanism is shown in Fig. 1, where *Cell* means a long chain of β-glucose units [33–36]. At the beginning of hydrolysis, the glycosidic oxygen is quickly protonated. The rate-limiting step is the flexure of the glucose molecule from the chair to the semiplanar configuration, accompanied by the elimination of the *Cell* residue from the glucose unit. The next steps are the rapid addition of water and the fast regeneration of the proton. The rotational energy required in the ring flexure seems to be the rate-controlling factor in hydrolysis, and the slow hydrolysis rate of cellulose is explained by the rigidity of the glucose rings held tightly in the crystal structure determined by the hydrogen bonding between hydroxyl groups and hydrogen atoms of adjacent chains. The hydrolysis rates of amorphous cellulose and of hemicelluloses are much faster because the amorphous nature does not hinder the flexure of the ring. The hydrolysis of amorphous pentosans and hexosans can thus be carried out at low temperature and low acid concentration (e.g. 1% concentration and 150 °C) but to break the chemically resistant crystalline cellulose it is necessary to use either dilute acid with temperatures in excess of 180 °C, or to use concentrated acids at lower temperatures.

While the hydrolysis of pentosans is rapid, they are also easily decomposed to furfural and tars. Table 2 shows the values of the rate constants, in minutes, at two

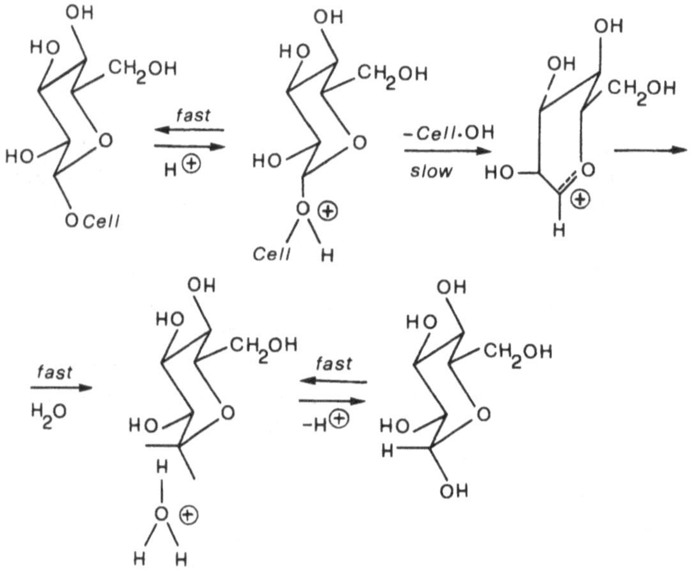

Fig. 1. Hydrolysis of cellulose via the cyclic carbonium-oxide ion [from [37]]

different acid concentrations and at various temperatures [33–36] assuming that hydrolysis and decomposition are modeled as first-order homogeneous reactions. The table shows that higher temperatures can provide slightly improved ratios of hydrolysis to decomposition.

Table 3 (same references and assumptions as for Table 2) lists the rate constants for the hydrolysis of cellulose and the glucose decomposition to hydroxymethylfurfural with dilute sulfuric acid. It can be seen that, theoretically, the maximum advantage,

Table 2. Rate constants of hydrolysis of xylans and of xylose decomposition (K in min^{-1})

	H_2SO_4 1%			H_2SO_4 0.6%		
Temp. °C	K_{xyl}	K_{deg}	K_{xyl}/K_{deg}	K_{xyl}	K_{deg}	K_{xyl}/K_{deg}
120	2.6×10^{-2}	2.4×10^{-3}	10.8	1.4×10^{-2}	1.7×10^{-3}	8.2
150	3.2×10^{-1}	2.7×10^{-2}	11.8	1.8×10^{-1}	1.9×10^{-2}	9.5
180	2.8	2.3×10^{-1}	12.2	1.6	1.6×10^{-1}	10.0

Table 3. Rate constants of hydrolysis of cellulose and of glucose decomposition (K in min^{-1})

	H_2SO_4 1%			H_2SO_4 0.6%		
Temp. °C	K_{glu}	K_{deg}	K_{glu}/K_{deg}	K_{glu}	K_{deg}	K_{glu}/K_{deg}
180	1.02×10^{-1}	3.34×10^{-1}	0.305	6.0×10^{-2}	2.49×10^{-1}	0.241
210	1.30	1.40	0.929	7.6×10^{-1}	1.05	0.723
240	12.10	5.00	2.42	7.14	3.73	1.91

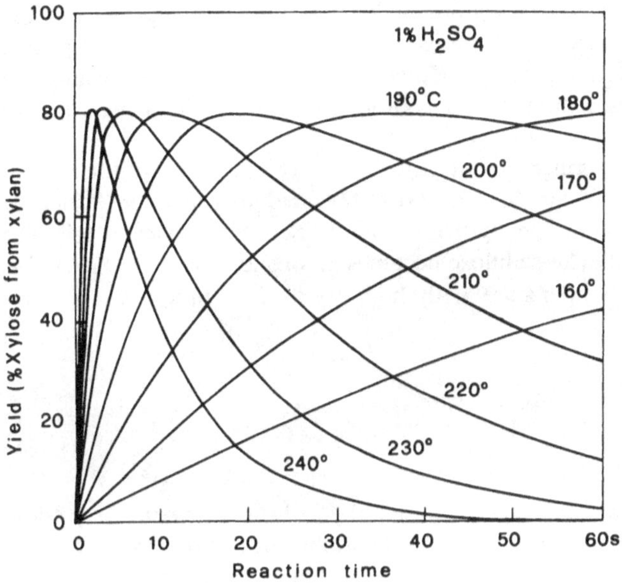

Fig. 2. Xylose yield from the acid hydrolysis of xylan as a function of reaction time and temperature [from [37]]

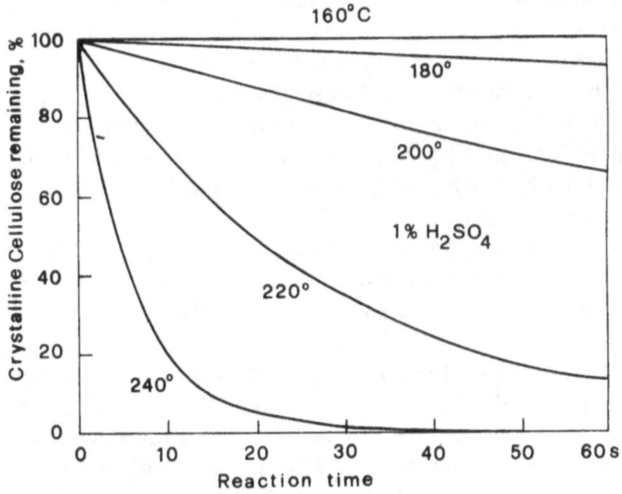

Fig. 3. Acid hydrolysis of cellulose as a function of reaction time and temperature [from [37]]

corresponding to the maximum value of the ratio K_{glu}/K_{deg}, is reached at the highest temperatures and acid concentrations.

The yield in xylose from xylosan and the yield in glucose from cellulose are given in Figs. 2 and 3 respectively, at different times and for different temperatures with an acid concentration of 1% [37].

As can be seen, high yields in dilute acid hydrolysis can be achieved either with high temperature and/or acid concentration, or by removing the sugars from the reactor before they have a chance to decompose. This last condition may be realized, for instance, by using percolation reactors. However, the water used in washing the sugars from the reactor causes dilute sugar solutions, which result in high capital and energy costs.

In cellulose hydrolysis by concentrated acids (i.e. concentrated sulfuric acid, fuming or gaseous HCl, liquid or gaseous HF), the acid disrupts the lattice of crystalline cellulose by breaking the hydrogen bonds between adjacent cellulose chains, with evolution of heat. The cellulose becomes amorphous in character and is easily hydrolyzable at low temperature with high yields and practically no by-products.

3.3 The State of the Art

Comparative studies of the various systems for acid hydrolysis have compared these in terms of yield and costs. These studies show the importance of maximizing yield, and minimizing the formation of sugar degradation products, which may inhibit the bioconversion of sugars to ethanol [38]. In order to accomplish this, one can use strong acids at moderate temperature, or dilute sulfuric acid at high temperatures or in percolation type reactors.

With the choice of a strong acid, the problem of its recovery is paramount. The original Bergius process (with concentrated HCl) has given rise to a series of possible improvements, aimed at making the process continuous [39] and at making the acid recovery easier [40,41]. The process, however, is still basically the same: the acid is fed in at 41%, exits from the hydrolysis at about 30%, and must then be reconcentrated. The hydrolysis takes place at about 35 °C for 1 h, with a yield of essentially 100% of biodegradable sugars. Post-hydrolysis with diluted acid may be necessary to complete the hydrolysis of some oligomers. The high cost of the acid recovery system in the liquid-phase HCl hydrolysis process is caused by the high volume of liquid to be processed, and the extremely corrosive nature of HCl, which requires the use of expensive alloys as Hastelloy or Monel, glass-lined steel, and graphite for the heat exchangers.

In order to reduce acid recovery costs, the use of gaseous HCl has been suggested [42-44]. For such a process, wood chips dried to less than 10% moisture are impregnated with gaseous HCl at low temperature. The heat is removed by circulating cold HCl from a self-refrigerating expansion system. The impregnation is carried out at high pressure (2000 to 2200 kPa) in a fluidized bed. The self-refrigeration system acts through compression of the HCl at the previously stated pressure and subsequent expansion. Although the cost of such a plant is only two thirds of the cost of a liquid phase HCl system, this is offset by higher losses of acid. No plant adopting this system has been constructed on an industrial scale.

The use of HF in liquid [31,32,45-49] or gaseous form [50,51] would also have a yield of essentially 100%. However, because hydrofluoric acid costs five times as much as hydrochloric acid, even small losses are economically significant. Losses will occur due to the difficulty of desorbing HF from the carbohydrate produced and by loss as an HF/H_2O azeotrope in the recovery section. Moreover, even small losses from the plant could be extremely dangerous because of the high toxicity of the acid.

The temperature in the hydrolyzer is about 23 °C and the ratio of HF to lignocellulosic is 2.7 to 1 by weight. The reactor can be carbon steel. Duration of hydrolysis is about 20 min. HF is recovered by repeatedly heating and flashing of the products. The post-hydrolysis, which could be theoretically realized with the $HF-H_2O$ azeotrope, must in fact use diluted sulfuric acid to avoid the presence of the fluoride ion in the hydrolyzate, since this ion is toxic to most microorganisms. The cost of the plant, which may at first glance seem comparatively modest because of the low cost materials employed, is actually much higher because of the requirement for perfect containment of the HF and because a refrigeration plant needs to remove the heat of reaction.

The use of concentrated sulfuric acid [52-59], which is considerably less expensive than hydrochloric and hydrofluoric acids ($ 0.09 kg^{-1} vs $ 0.26 and $ 1.35 respectively) makes recovery less important. The acid is neutralized with lime (a demonstration recovery plant working with membranes in Japan revealed the cumbersomeness of the process). Still, the gypsum formed during neutralization is formed in great quantities (e. g. twice the weight of the produced ethanol), and the cost of the acid employed and the lime necessary to neutralize it ends up being not less than that expected for a plant based on liquid HCl, but exceeds it by almost 50%. However, the capital investment is considerably less than that of a liquid HCl process (half as much) because the acid can be handled in fiberglass and plastic equipment. The processing scheme, as realized in a pilot plant of the Tennessee Valley Authority (Muscle Shoals, Alabama) [60,61] is shown in Fig. 4. It is worth noting, in particular, that prehydrolysis of hemicelluloses is carried out with the acidified product from the cellulose hydrolysis. Solids, once washed, centrifuged, and dried at 85 C to 10% moisture, are mixed with sulfuric acid and the hydrolysis of cellulose is carried out

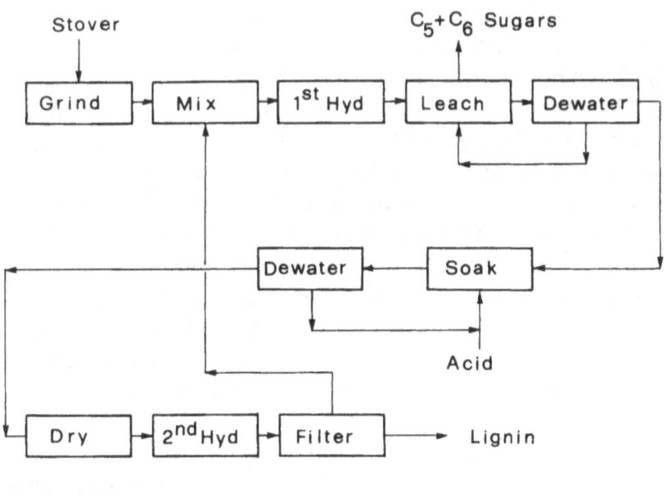

Fig. 4. Simplified scheme of hydrolysis of lignocellulosics with concentrated sulfuric acid (Tennessee Valley Authority)

in 30 min at 140 °C. This scheme minimizes the formation of inhibitors from hemicelluloses.

Alternative processing schemes using high temperature dilute sulfuric acid and various plug flow designs have been advanced which require acid concentrations of 0.5 to 1.5% and temperatures between 180 and 240 °C [62-65]. Hydrolysis durations range from a few minutes to a few seconds, in inverse relation to temperature. If the reaction is carried out in a single, high temperature step, the pentose sugars are efficiently converted to furfural. However, glucose yields do not exceed 55% and large amounts of degradation products are formed, which inhibit the bioconversion to ethanol. The process, though, has the economic advantage of treating relatively high concentrations of lignocellulosics and thus obtaining hydrolyzates with comparatively high glucose concentration [66], although there are important mechanical problems in moving mixtures of lignocellulosics and sulfuric acid with more than 20% dry matter.

Saving the xylose for a subsequent transformation requires the process to be divided into two steps. The first step is the prehydrolysis of hemicelluloses by dilute sulfuric acid at moderate temperatures or by acetic acid (1%) or calcium phosphate at 205 °C [67]. The second step consists of the conditions normally adopted for a simplified plug flow process [67-73].

For several years now, just because of problems caused by other processes, many proposals have been made for improvements of the Scholler process to improve yields and sugar concentrations [55,74-81]. The acid concentration has been raised to 2% or even 4%, and the stages of hydrolysis of hemicelluloses and of cellulose have been separated, each to be performed at different temperatures (140 to 160 °C and 160 to 180 °C and over, respectively). Correspondingly, residence times have been reduced: for instance, from the 3.5 h of the original Scholler process to 20 to 90 min. In this way, yield rises to 70 to 80% and glucose concentration may reach 12 to 14% if a suitable recirculation of the hydrolyzate is performed. As a general rule, the trend is to keep the hydrolyzates of hemicelluloses separated from those of cellulose, either with the purpose of using the xylose in a different way, or because the bioconversion of glucose and xylose to ethanol (or to any other product) may be best carried out separately. A remedy to the aggressivity of dilute sulfuric acid at high temperatures is the use of Monel linings for the prehydrolysis and the use of Monel or Zircalloy or tantalum linings for the hydrolysis stage. Shortening residence times obviously reduces the reactors volume and, consequently, lowers their cost.

A particularly interesting process based on the Scholler process that is under study is the progressing-batch dilute-acid hydrolysis process, which combines the simplicity of operation of the Scholler process with the advantages of the counter-current operation [82-83]. The continuous counter-current reactor is attractive on paper, but when using lignocellulosics, it is a very complicated operation. One solution is to simulate a continuous process by means of several batch reactors in series, as shown in Fig. 5. The liquid enters reactor 6, proceeds towards the left, and exits from reactor 2. The hydrolyzate is flashed to quench the hydrolysis and degradation reactions. After 15 to 30 min, the reactor stages are changed: fresh liquid enters reactor 5 and exits from reactor 1, reactor 7 is filled with fresh lignocellulosic, and residual solids in reactor 6 are discharged. The net effect is that fresh solids enter the reactor train at the left and spent solids are discharged

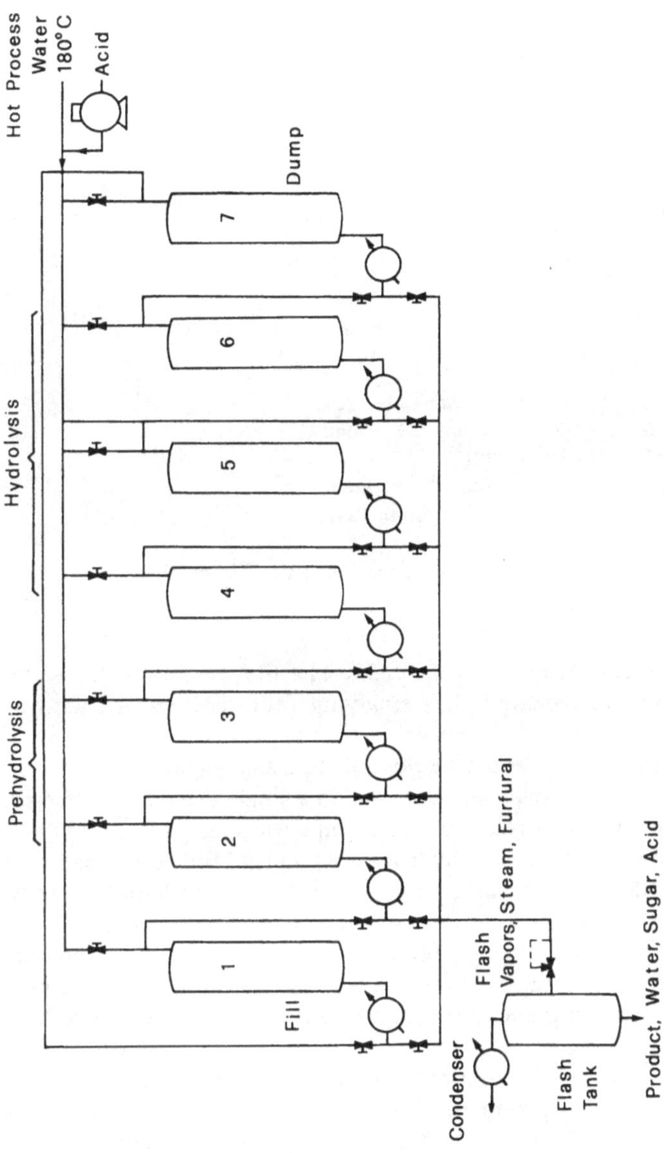

Fig. 5. Progressing batch reactor schematic [from [82]]

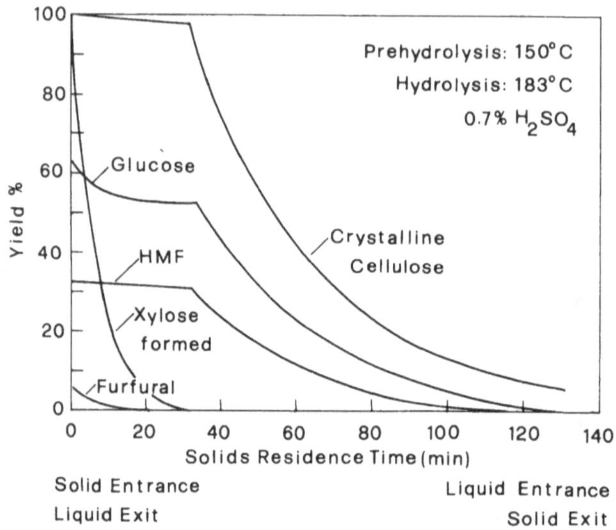

Fig. 6. Progressing batch reactor. Yields as a function of residence time [from [82]]

after leaving the last hydrolysis reactor on the right. The first reactor in the series is used as à prehydrolysis reactor, and the remaining four reactors are used as hydrolysis reactors.

It is calculated that in this way it should be possible to attain sugar yields of 80% and to increase concentration 20% over that obtained in a single percolation reactor. The design conditions for the reactor system are a solid residence time of 130 min and a liquid residence time of 45 min (prehydrolysis at 150 °C for 30 minutes and hydrolysis at 183 °C for 100 min) with an acid concentration in the liquid phase of 0.7%. Roughly, yields as a function of time are expected as reported in Fig. 6. A true counter-current process is being studied [81], using a reactor commonly employed by the pulp and paper industry, in which the material fills a chain of small basins moving counter-current to the acid solution. No final results are available at present.

Various techniques [84] to overcome inhibitor formation in acid hydrolysis processes and to increase yields have been proposed. On toxicity for yeast in aerobic and anaerobic conditions, see also Lüers et al. [85]. If acid hydrolysis is performed in separate stages, the first stage contains more pentoses and fewer hexoses; the second stage contains more hexoses and few pentoses. Despite the presence of pentoses, for ethanol production no special techniques are generally required for the hydrolyzate of the second stage, which is usually biodegraded directly by *S. cerevisiae*, *S. uvarum*, or *Zymomonas mobilis*.

According to Beck and Strickland [80,86], who have exhaustively researched the subject, a warm treatment with $Ca(OH)_2$ at ph 5.5 with the addition of sulphite, or an overliming at pH 10, allows better yields in ethanol production, particularly in the presence of hemicellulose hydrolyzates. However, designing clean-up treatments to prepare the hydrolyzates for biotransformation is poorly understood, and is largely a matter of trial and error. Purification procedures may be even more necessary when processes with immobilized cells [80,87,88] or loop bioreactors are used.

4 Enzymatic Hydrolysis

4.1 The *Trichoderma* Route

Enzymatic hydrolysis has very high yields, because cellulase enzymes only catalyze hydrolysis reactions, and not sugar degradation reactions. However, relevant research has been conducted mainly over the past decade, and considerable improvement is necessary before it is economically competitive. The most important research issues are: improving the performance and understanding of pretreatment; producing less expensive and more effective enzymes, and developing hydrolysis processes with greater yields, product concentrations, and rates.

In 1950, E. T. Reese isolated and described [89] a fungus, *Trichoderma viride*, subsequently classified as *T. reesei*, capable of producing a suitable active cellulase. Many articles published in this series [1–5,7–10], concerned in particular with the nature of the enzyme and the kinetics of enzymatic hydrolysis, should be consulted for more detailed information. As the literature concerned has already reached several thousand papers, it is beyond the scope of this work to supply a comprehensive list of references that would anyway tend to be incomplete. Reference is made here only to publications and results that appear to be most pertinent.

The enzyme cellulase system is a mixture of *endo*-β-1,4-glucanglucanhydrolases (EC 3,2,1,4), *exo*-β-1,4-glucancellobiohydrolases (EC 3,2,1,91), and a β-glucosidase (EC 3,2,1,21). The enzyme system has been found in numerous mesophilic and thermophilic fungi, mesophilic and thermophilic bacteria, and some *Actinomycetes*. The first enzyme acts randomly on the interior of the cellulose polymer to generate new chain ends and is competitively inhibited by end products. The second enzyme catalyzes the cleavage of a cellobiose unit from the non-reducing end of the cellulose chain and appears to be end-product inhibited. The last component splits cellobiose and other oligomers to glucose and is also end-product inhibited by a non-competitive mechanism [3,90]. Several microorganisms produce also other hydrolytic enzymes such as *endo*-1,4-β-xylanase, β-xylosidase, α-1,3-arabinosidase, α-1,2-glucuronosidase, mannanase and acetylesterase [91]. Spray-drying methods for cellulase preservation have been studied: xylanase is destroyed and β-glucosidase is largely inactivated by such a process [92].

The composite nature of cellulase and the inhibition phenomena connected to it explain one of the major trends of present research on enzymatic hydrolysis of cellulose. Using the wild strain of *Trichoderma*, glucose product concentrations were limited to 2.0–2.5%. Such a concentration could be acceptable for acetone-butanol, but not for industrial ethanol production, as the cost of recovery of ethanol from a 1% medium is prohibitive and the product inhibition from ethanol is well over 1%. Clearly, the most interesting possibility was to produce a mutant *Trichoderma* that could supply a high quantity of β-glucosidase. Research work with this objective has been carried out, particularly in the USA at Rutgers University, the US Army Natick Laboratories (MA) and Genencor; as well as in Europe, in Finland [93], in

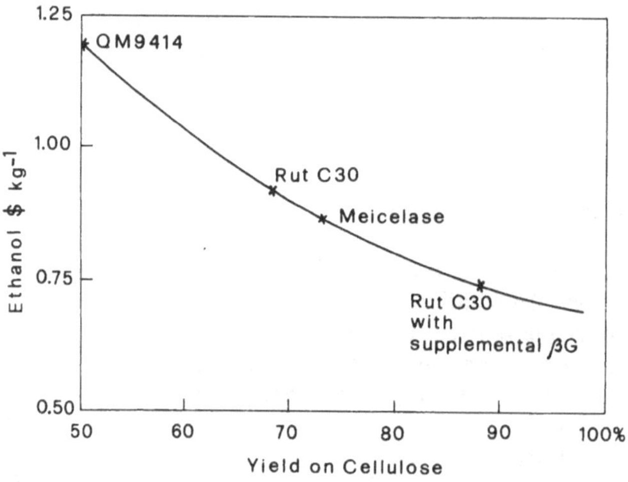

Fig. 7. Cost of ethanol if different enzymes are used for cellulose hydrolysis [from [97]]

France and Denmark (Novo Industri). The productivity of cellulase producing fungi has been increased by almost two orders of magnitude.

At the same time attempts have been made to increase the specific activity of the system. The activity is a function of enzyme and substrate composition. As a general rule, an I.U. is the quantity of enzyme necessary to produce glucose at a rate of 1 μmol min^{-1}. The activity of endoglucanase is usually measured on carboxymethylcellulose and the activity of glucosidase on cellobiose. The most prevalent assay of the total system is to measure its activity against filter paper (in this case, one inserts in I.U. the letters F.P.). The specific activity of cellulase has been increased by only a factor of two.

The addition of β-glucosidase produced from other sources, e.g. *Aspergillus niger* or *wentii* [94–96], has been tested. Figure 7 shows the variation of ethanol cost using different enzymes and an enzyme added with β-glucosidase. The final concentration when β-glucosidase has been added may reach 8% of glucose versus 4.3%, which is attainable, for example, with the enzyme Rut C30 alone [97].

β-Glucosidase is also subject to product inhibition; therefore an excess of it is not as useful as the continuous removal of glucose, e.g. by bioconversion [98]. Inhibition from ethanol can be considered less important than that from glucose, even at equal concentration. Processes in which the hydrolysis and alcoholic "fermentation" are carried out simultaneously are referred as "simultaneous saccharification and fermentation" (SSF) or "combined hydrolysis and fermentation" (CHF).

Because the optimum temperature for cellulase ranges from 45 to 50 °C, while that of *S. cerevisiae* from 30 to 32 °C, a large number of tests have been carried out with various microorganisms as *Schyzosaccharomyces pombe* [98], *Candida brassicae* [98–101], *Candida lusitaniae* [102,103], *Candida acidothermophilum* [104], *Zymomonas mobilis* [105,106], a thermoresistant mutant of *Saccharomyces cerevisiae* [103,108], and *Brettanomyces* spp. [107–109]. The most recent conclusions [109] seem to indicate that the most thermotolerant microorganisms present greater process difficulties, while several organisms perform well at 36 to 40 °C. *Brettanomyces clausenii*, which has to be used with the thermotolerant mutants of *S. cerevisiae*, allows cellobiose to be exploited directly, further reducing inhibition.

The production of the enzyme presents several problems. Production on a suitable lignocellulosic substrate is an aerobic process, the productivity of which is around 50 IU L^{-1} h^{-1}. After a residence time of 13 d, a titre of 15 IU mL^{-1} is reached. The long residence time causes high capital costs [97] and increases the risk of bioreactor contamination. Hydrolysis requires 3 d, employing 15 to 20 IU of enzyme per g of solids. After this time, about 70 to 80% of the cellulose and 100% of the xylans are hydrolyzed [110]. Several attempts to speed up and increase the enzyme production have been made, using, for instance, lactose as a carbon source in a fed-batch mode under permanent sugar limitation. Under these conditions it is possible to arrive at 22 IU mL^{-1} [93, 111, 112]. Lactose (of whey) is also a cheaper carbohydrate than relatively pure cellulose (see Sect. 4.4). However, cellulose, possibly through its hydrolysis intermediates, should normally be added as an inducer [113–115].

4.2 Other Cellulolytic Microorganisms

Numerous microorganisms have been tested for their ability to produce cellulase. With no intention to be exhaustive, mention will be made here of fungi such as *Chrysosporium lignorum* [116], of *Fusarium solani* [117–120] and *F. oxysporum* [121–123], of *Neocallimastix frontalis* [124], of *Penicillium funiculosum* [125] and *P. pinophilum* [126, 127], of *Chaetomium cellulolyticum* [113–115, 128], and, among thermophilic fungi, *Tularomyces* spp. [129–131], and *Sporothricum cellulolyticum* and *S. cellulophilum* [132]. Among mesophilic bacteria, *Acetovibrio cellulolyticus* and *Clostridium cellulolyticum* [133, 134]; among thermophilic bacteria, *Clostridium thermocellum* [135–139]; among *Actynomycetes*, *Micromonospora calcae*, *Pseudonocardia thermophila* and *Thermomonospora* spp. [140–143].

Cl. thermocellum is now attracting the most interest. According to Halliwell et al. [144], this bacterium is endowed with the most powerful cellulasic system. Intensive genetic engineering work is under way at the Massachusetts Institute of Technology and elsewhere [139, 144–146]. This *Clostridium* works at about 60 °C and is also capable of converting glucose and cellobiose (but not pentoses) to ethanol. Therefore, a co-culture with *Cl. saccharolyticum* has been suggested in order to use the pentoses simultaneously. The co-culture speeds up the process by eliminating oligosaccharides: after 48 h, the conversion percentage is approximately three times higher than in a single culture of *Cl. thermocellum* [147]. However, the co-culture is not devoid of inconveniences, since ethanol, acetic, lactic, and butyric acids form simultaneously. Some authors have, thus, suggested the co-culture with *Cl. thermohydrosulphuricum* [138], which does not produce isobutyric acid, or with *Thermoanaerobacter ethanolicus* and its mutants [121]. The reason for resorting to these thermophilic microorganisms is to take the maximum advantage of the thermophilic cellulase from *Cl. thermocellum*.

A thermophilic fungus, *Talaromyces emersonii*, is the object of intensive research because it is apparently capable of producing a high quantity of β-glucosidase when cultivated on lactose. Similarly, work is being carried on *Fusarium oxysporum*, which is, however, mesophilic and very sensitive to the inhibitors that may form during a pretreatment by steam explosion. This fungus supplies a high percentage of xylitol with ethanol [122]. Two other recently isolated fungi are *Penicillium pinophilum*, a mutant of which seems able to give in culture 10 I.U. mL^{-1}, but with a high β-glucosidase content [126], and secondly *Neocallimastix frontalis*, which has yielded (in 72 h and with an enzyme concentration of 0.370 IU mL^{-1}) 100% of solubilization versus the 38% given by *T. reesei* Rut C-30 under the same conditions [124].

The *Actynomycetes* group is an interesting group of cellulolytic microorganisms, which are easier to grow than fungi and whose strain improvement by the application of recombinant DNA technology is more tenable. Work is under way, as the most recent references reported above show, but no satisfactory conclusions have been reached to date.

For the sake of completeness, we must mention an attempt to transform enzymatic hydrolysis from a heterogeneous catalytic process into a homogeneous process. This is realized by transforming cellulose into acetylcellulose (16.2 to 16.4% of acetyl groups) and attacking it with the enzyme of *Pestalotiopsis westerdijkii*. Enzyme production from this fungus is completed in 5–8 d; the attack lasts 3 to 5 hours and can be followed by an inoculum with *S. cerevisiae*. Only 0.7 g of enzyme per kg of acetylcellulose is required, versus 15 g of enzyme from *T. reesei* per kg of cellulose [148]. No comparative evaluation of this process has been made.

4.3 The Pretreatment

Lignocellulosics are rather resistant to enzymatic hydrolysis unless a suitable pretreatment is used. The surface area available for the enzyme–substrate interaction will be influenced by pore size and shielding effect by hemicelluloses and lignin. The crystalline structure excludes water molecules as well as any larger molecule and thus reduces available surface area. For hydrolysis of cellulose to occur, the enzyme must bind to the surface of the cellulose molecule to catalyze the reaction. Cellulase enzymes have a molecular weight of 30 to 60 kDa and an ellipsoidal shape with major and minor dimensions of roughly 30 and 200 Å [149,150]. Consequently, only 20% of the pore volume is accessible to cellulase molecules, and the number of accessible pores can be increased if hemicellulose and lignin are removed.

Pretreatment can be physical or chemical. Physical pretreatments include various forms of milling, shredding, and mulching. Chemical pretreatments enhance enzymatic susceptibility by removing the shielding effect of lignin, reducing crystallinity, and increasing cellulose solubility or swelling. Many of these processes have been examined in former volumes of this series [6,9,10]. At present, the three pretreatment processes that attract most attention are organosolv, which takes advantage of the dissolving properties of water-alcohol (methanol, ethanol) mixtures, steam treatment at 180 to 240 °C for 1 to 30 min [151–155], and dilute acid prehydrolysis.

Steam-explosion is widely described in the literature [153–155]. It should be pointed out that the vitreous transition temperature is approximately 125 °C for lignin, 165 °C for xylans, and 234 °C for cellulose. It is important to reach this last

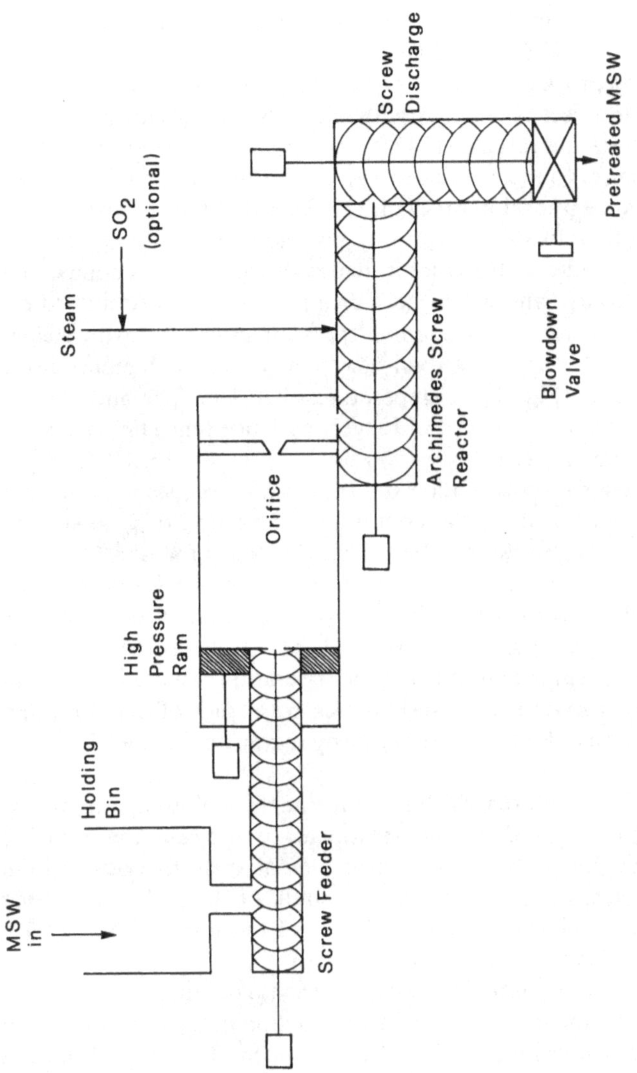

Fig. 8. The stake pretreatment process (schematic)

temperature in a short time (45 to 60 s at most) to avoid decomposition phenomena that become very evident at 260 °C for cellulose as well.

The Iotech steam-explosion process is a batch process very similar to that for masonite production. An explosion gun is charged with biomass, and steam is injected until the desired cook temperature is attained (245 to 250 °C). After the required cooking time (about 5 s), the pressure is quickly reduced, explosively discharging the content in a cyclone. Some water and degradation products are flashed in the cyclone.

The Stake process is a continuous steam-explosion process in which the lignocellulosic is transferred by a screwfeeder to the reactor, where the feedstock is compacted by a high-pressure ram through an orifice and dropped onto an Archimedean screw, where high pressure and temperature come from direct contact with steam. The letdown is accomplished through a ball valve, which causes the lignocellulosic to explode into fine particles (Fig. 8). Temperatures between 220 and 235 °C are recommended [156]. Lower temperatures and longer residence time (190 to 200 °C and 10 min.) are also being investigated.

Recents studies [247, 248] have shown that the explosion is not necessary to increase hydrolysis rates. On this ground and on the continuous extraction of gaseous and soluble products is based the RApid Steam Hydrolysis (RASH) process [249, 250].

Steam treatments release uronic acids and acetyl groups in the form of acetic acid, which carry out a hydrolysis of hemicelluloses. A complete hydrolysis can be achieved by introducing small quantities of mineral acids or SO_2. Weak prehydrolysis pretreatment, whatever the chemical or mechanical operation, is useful for two reasons: it separates the hydrolyzate of hemicelluloses from that of cellulose and even increases the surface area available to the enzyme by removing the shielding layer of hemicelluloses [157].

Dilute acid pretreatment is also a prehydrolysis of the hemicellulose-lignin matrix. In this process, dilute acid is added to the treatment to increase the ratio of xylan hydrolysis to xylose degradation. While conversions of xylan to xylose are on the order of 30 to 50% in steam treatments (with the bulk of the scylan converted to furfural and degradation products) xylose yields are of the order of 80 to 90% for systems in which dilute sulfuric acid is used [157–160].

Organosolv pretreatment adds an organic solvent (methanol or ethanol) to the pretreatment reactor to dissolve and remove the lignin fraction. In the pretreatment reactor the internal lignin-hemicellulose bonds are broken and both fractions solubilized, while cellulose remains as a solid. Many combinations of solvent concentration, acid type and concentration, temperature and time are possible. After leaving the reactor, the organic fraction is removed by evaporation and recycled to the reactor. Without an organic fraction in the liquid phase, lignin precipitates and can be removed by filtration or centrifugation. Thus this process separates the feedstock into a solid cellulose residue which is easily digestible, a solid lignin which has undergone few condensation reactions, and a liquid stream containing xylose [161].

Another process, based on treatment with 12% NaOH for 4 h at 80 °C is suggested by Cunningham [162]. Other processes adopted by the pulp and paper industry are not suitable for commercial application in this case. Processes for enzymatic solubilization of lignin [19, 163] are also very interesting from a scientific point of view but are actually too expensive for commercial application.

4.4 Enzyme Production and Hydrolysis Processes

Here we shall consider *T. reesei* because more realistic data are available and it has potential for improvement.

As previously mentioned, enzyme production takes approximately 13 days and represents the single most costly section of the plant. The best results are obtained in a fed-batch system [150] where the substrate is slowly added to the bioreactor, reaching a final loading of 150 g of biomass per litre. The inoculum is prepared in a first-stage bioreactor having a working volume of 10% of the volume of the production stage, which has a working volume of 10% of the daily production. At the end, the final mixture is filtered through a rotatory filter and stored. The remaining solids (about 18 g L^{-1} of mycelium and 7.5 g L^{-1} of cellulose) are sent to byproduct recovery (as animal feed) or burnt. Given these conditions up to 100 IU L^{-1} h^{-1} can be obtained on steam-exploded agricultural residues and 50 IU on aspen wood. It is also reported in the literature [111] that it is possible to raise productivity to 400 IU L^{-1} h^{-1} with a particular mutant of *T. reesei* and lactose as the carbon source, but with steam-exploded aspen wood or rayon cellulose as inducer. It seems now, that it is possible to use lactose alone and that it could be possible to use the hemicelluloses hydrolyzate as 80% of the carbon source together with 20% of rayon pulp as inducer [164].

The production of cellulase by an SSF ("solid-state fermentation") [165] and a recycle process with a two phase system and ultrafiltration [166] have been also proposed.

The most important parameters in the hydrolysis section (yield, concentration, hydrolysis duration, and required enzyme loading) are strictly interrelated, once the pretreatment and the nature of the enzyme are selected. Yields are higher in more dilute systems, where inhibition is minimized and where increasing the enzyme loading can, to a limited extent, overcome inhibition and increase yield and product concentration. Finally, longer reaction times also make higher yield and concentration possible. However, this is not a simple optimization problem.

When hydrolysis and biological conversion are conducted separately, hydrolysis is carried out at 45 °C and pH 5. By adding fresh feed over a number of hours after the hydrolysis begins, high substrate loadings can be achieved and mixing problems avoided. Enzyme loadings are in the range of 20 to 50 IU g^{-1} of cellulose. Figure 9 [97] shows the effect of enzyme loading on the cost of a final product such as ethanol. From the curve, given the high cost of the enzyme, higher savings are attained with enzyme loadings of approximately 12 IU g^{-1} of cellulose. After this point, it appears that the available sites on the cellulose surface are practically saturated, and adding more enzyme (e.g. the 20 IU quoted before) is useful only because it provides more β-glucosidase.

Recovery of the the enzyme is one method of minimizing the effect of the high enzyme cost. Physical separation methods such as ultrafiltration are impractical because of high costs [167]. The most promising techniques make use of the high affinity of cellulase for cellulose. This can be exploited by a process of countercurrent absorption of the *endo*- and *exo*-enzyme components released at the end of the hydrolysis on fresh feed. As a consequence, the recovered enzyme still needs to be supplemented with β-glucosidase. Also, a fraction of the enzyme is degraded during hydrolysis and another part adheres to lignin and non-hydrolyzed cellulose. The

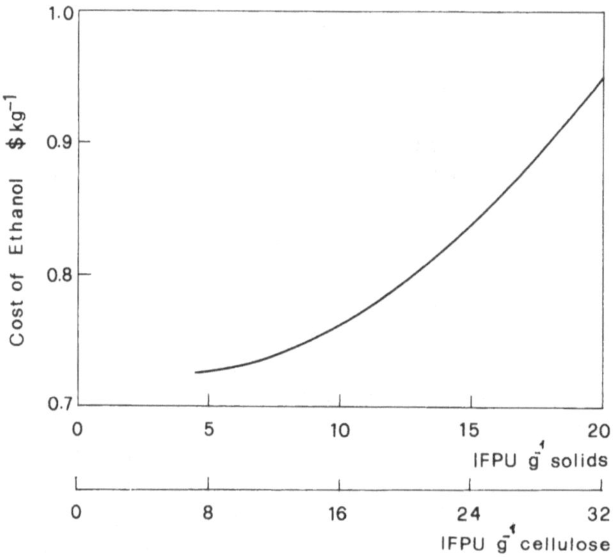

Fig. 9. Effect of enzyme loading on the cost of ethanol [from [97]]

major problem with such processes is that they greatly increase the risk of contamination. Immobilization of the β-glucosidase would reduce the amount which needs to be made up [168–171].

Energy consumption of mixing during enzyme production and hydrolysis has not been widely discussed in the literature. Mixing to enhance oxygen transfer during enzyme production is one of the major costs; mixing requirements during hydrolysis are not well defined, but it appears that the major function of such stirring is to prevent high local super concentrations which inhibit enzyme activity. The power input per unit volume in hydrolysis reactors could be quite low.

If ethanol is the desired final product, simultaneous saccharification and bioconversion ("fermentation") SSF process can alleviate many of the limitations of enzymatic hydrolysis. SSF processes require less enzyme (7 to 15 IU g^{-1} of cellulose) because inhibition is reduced. Also final product concentrations are higher (4% ethanol), and yields are also improved.

5 The Utilization of Hydrolyzates and By-products

5.1 Ethanol Production

Problems connected with ethanol production from cellulose and various hydrolyzates were quoted and discussed in the preceding sections and the more pertinent literature was quoted too. The treatment of the subject will not be repeated or summarized here. However, see also Sect. 6.

5.1.1 Ethanol from Pentoses

Biodegradation of a medium rich in pentoses (as in the first stage hydrolyzate of an acid hydrolysis process, or the hydrolyzate from a pretreatment process before enzymatic hydrolysis) is performed either with microorganisms capable of using pentoses directly (e.g. *Pachysolen tannophilus*, *Pichia stipitis*, *Candida shehatae*, *Clostridium thermosaccharolyticum*) or by first transforming xylose into xylulose using xylulose-isomerase and then *S. cerevisiae*.

P. tannophilus has been studied in depth [15,16,167,172-177]; in the presence of relatively modest quantities of glucose it degrades the glucose rapidly and the xylose more slowly, as shown in Fig. 10 [80]. However, appreciable quantities of xylitol are simultaneously formed. In anaerobic conditions, from five molecules of xylose, four molecules of xylitol and one of ethanol are formed; in partially aerobic conditions (0.025 vvm), 30 g L^{-1} of xylose yield 10 g L^{-1} of ethanol and 9 g L^{-1} of xylitol [90].

The slow utilization of xylose in anaerobic conditions, at least for ethanol production, results from an inbalance in production and consumption of NADH in the overall conversion of xylose to ethanol. For this reason, Alexander [178], by adding 100 to 200 mM L^{-1} of acetone, increased ethanol yields from xylose from 29.0 to between 38.7 and 45.0% by weight, respectively. Yields in xylitol were correspondingly reduced from 19.0 to between 16.5 and 6.5%.

An observation by Beck et al. [80] concerning *P. tannophilus* is particularly interesting. When glucose and xylose are present in equivalent amounts (as when hydrolysis is performed in a single stage), glucose is consumed in preference to xylose; xylose and ethanol are then consumed simultaneously (see Fig. 11). This confirms that it is worth keeping the hydrolyzates C_5 and C_6 separate.

More recent studies have considered *Pichia stipitis* and *Candida shehatae* for xylose utilisation [179-181]. Yields by weight on xylose, which are approximately 30% by weight with the original strains, attain values of 40 to 43% [181-183] and up to 46% [184] in 48 h with selected strains. These values are very close to the theoretical.

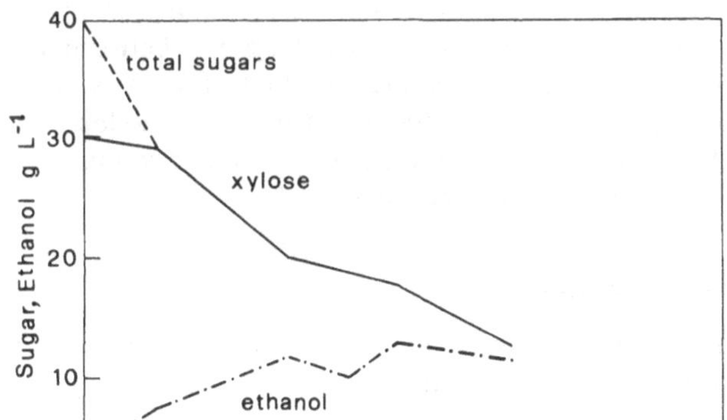

Fig. 10. Bioconversion to ethanol of sugars of a hemicellulose hydrolyzate with *Pachysolen tannophilus*. The C_5/C_6 sugars ratio is very high [80]

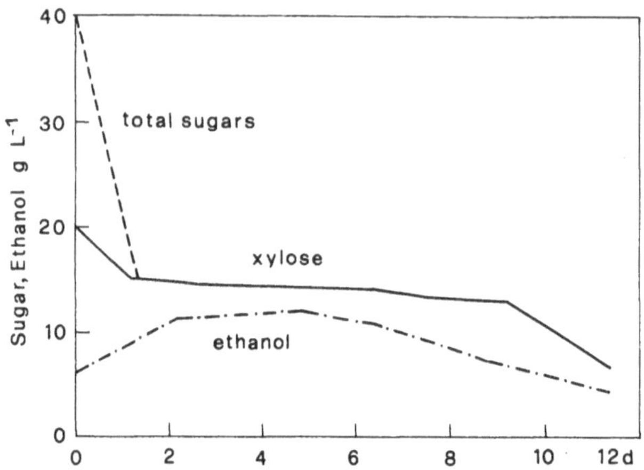

Fig. 11. Bioconversion to ethanol of sugars of a lignocellulosic hydrolyzate containing about the same quantity of C_5 and C_6 sugars with *Pachysolen tannophilus* [80']

These microorganisms suffer considerably from ethanol inhibition and maximum ethanol concentrations are of the order of 3% [185]. The two microorganisms only assimilate L-arabinose [186].

Recently, tests on continuous processes have been carried out as well [186-188]; comparison work has been performed by Grba et al. [189], Dellweg et al. [190] and Rizzi et al. [191,192].

Despite higher yields, *Fusarium oxysporum* appears not to be so interesting, because its working times extend to approximately 6 d [193,194].

Although common yeasts cannot directly give ethanol from xylose, many yeasts, including *S. cerevisiae*, can give ethanol from xylulose with high efficiency [195]. The major drawbacks of a process using xylose-isomerase to isomerize xylose to xylulose and then converting xylulose to ethanol with *S. cerevisiae*, is the high cost of the large quantity of enzyme required, and the difference in pH optima between the yeast (pH 5) and the enzyme (pH 7). Nevertheless a large amount of work is being carried out in this field, beginning with the overproduction of the *E. coli* xylose isomerase in *E. coli* and *B. subtilis* by DNA recombinant techniques and continuing with cloning and expression of the xylose isomerase gene in yeast [196-201].

5.2 Furfural Production and Utilization

Furfural is a byproduct of acid hydrolysis of pentosans and can be easily recovered and purified. If furfural were to be used as fuel in the plant, the purification phase could be eliminated.

Current uses of furfural are limited to that as a solvent (especially in the oil industry), to the production of plastic materials (urea-furfural), and to that of tetrahydrofuran and furfuryl alcohol.

Studies on other methods of furfural utilization are not new: numerous works by Reppe and coworkers cover this topic. Many more researchers searched for methods to obtain from furfural products such as butadiene, styrene, vinylfuran, adipic acid and adiponitrile, hexamethylenediamine, 1-butanol, lubricants and plastics. All these processes, some of which also had industrial application, have been abandoned because of the high cost of furfural. Should furfural become the by-product of very large-scale production, such as that of ethanol as a fuel, interest in its use could revive. For its uses, furfural should often be converted into furan (by elimination of the aldehyde group) and tetrahydrofuran (by subsequent catalytic hydrogenation).

From tetrahydrofuran it is possible, for instance, to produce 1,3-butadiene by dehydration-dehydrogenation [203]. About 1.8 t of furfural yield 1 t of butadiene, which is commonly used to produce acrylonitrile-butadiene-styrene (ABS) resins and styrene-butadiene (SBR) rubber.

Styrene, in turn, can be produced from 1,3-butadiene through a Diels-Alder reaction yielding 4-vinyl cyclohexene, which is subsequently passed to styrene [204]. Approximately 1.9 t of furfural give 1 t of styrene. Styrene can be employed to produce polystyrene in all its forms, ABS resins and SBR rubber. Attention could be given to 2-vinyl furan: the furanic homologue of styrene, although its synthesis still presents some difficulty in practice, could be certainly cheaper than styrene.

Also the production of adipic acid (and hexamethylenediamine) for nylon-66, through a Reppe synthesis [206,207], and of maleic anhydride [208], for which respectively 1 and 1.4 t of furfural per t are needed, could become interesting with low furfural costs.

5.3 Other Uses of the C_5 Hydrolyzates

Still in the field of C_5 hydrolyzates, the production of furanic polyols and xylitol from xylose is of interest. Furanic polyols are obtained from xylose by reaction in a hydro-alcoholic medium having active methylenic groups. They can be used both as intermediate products for fine chemicals and pharmaceutical chemistry and in the formulation of new polyurethanes presenting high thermal stability.

Xylitol is a non-cariogenic sweetener, the metabolism of which does not require insulin. However, for this last characteristic, the world consumption is not expected to exceed 100000 t a^{-1} [209]. Xylitol is obtained by catalytic hydrogenation of the xylose sirup and by its subsequent purification. The microbial reduction encounters less favour; yet it could be joined with ethanol production (see Sect. 5.1.1).

5.4 The Utilization of Lignin

For the production of ethanol from lignocellulosics to be economically competitive, a large part of the lignins should be used in higher value applications than as process fuel.

Much of the experimental work for the chemical utilization of lignin has been performed with lignins from the paper and pulp industry. Lignin from hydrolytic processes can be very different, depending on the process chosen. The best lignins may be those obtained through steam explosion or organosolv pretreatment for enzymatic hydrolysis [210]. The best lignins from acid processes are definitely those from halogen acid processes.

For many applications, see in this series the exhaustive paper by Janshekar and Fiechter [19].

The production of vanillin and syringic aldehyde by alkaline hydrolysis and oxydation with air is well known, but syringic aldehyde has no commercial applications and the vanillin market is not so large compared to the fuel-ethanol market.

The hydrocracking and dealkylation of lignins give a 38% yield of monophenols and a 8% yield of higher substituted phenolic compounds and catechol [211-213]. After dealkylation, the monophenols yield 24% phenol, 13% benzene, 22.5% light, mostly gaseous hydrocarbons, and 22% of a heavy liquid [212]. According to Goheen [214], a hydrocracking process would be interesting only if the yield of phenols reached 50%. A different approach would be to convert the phenols into methyl aryl and methyl substituted aryl ethers instead of separating the different phenols. These ethers could be used with good results as gasoline extenders and octane enhancers [215].

Many other possible uses are described by Janshekar and Fiechter [19]. Here we intend to add only some complementary information.

An interesting suggestion by Hsu and Glaser [216-218] about plastics is that lignin can be reacted with maleic anhydride and then with propylene oxide, to obtain a lignin-polyester-polyether that can react with diisocyanates to give polyurethane foams, cast films, adhesives, and coatings. A negative aspect of the use of lignin for plastics is the color: a recent US patent [219] obtains lignin of a much less deep hue.

In addition, lignin can be used as a raw material for phenol-formaldehyde and urea-formaldehyde resin type resins and adhesives and also as an extender, mainly for plywood and laminates production [for a general review, see Nimz [220] and Gillespie [221]].

Comparisons between Kraft lignin and bioconversion lignin are made by Muller [222,223].

Because alcoholic and phenolic groups are simultaneously present, lignin can be ethoxylated to form a water-soluble polymer: the amount of ethoxylation and the choice of the lignin source (steam explosion, organosolv or acid hydrolysis) can give the desired degree of water solubility [224]. This could open to lignin the field of non-ionic surfactants, with some analogy with those derived from nonylphenol, which are much more valuable than anionic surfactants as lignosulfonates, even though they are marketed in smaller amounts.

Finally, an interesting use of lignin from hydrolysis processes could be that of the reinforcement of rubber in place of carbon black [225-227].

5.5 Other Utilizations of Hydrolyzates

Acetone-butanol production by anaerobic conversion of sugars is very well known. The same microorganism (*Clostridium acetobutylicum*) can utilize glucose and xylose and some work is still being done, particularly in France and in the U.S. to improve such a process. The concentration of resulting solvents (acetone-butanol-ethanol) cannot exceed 16 to 18 $g\,L^{-1}$ because of the inhibitory effect of butanol. Some genetic engineering work has allowed conversions to reach 23 to 24 $g\,L^{-1}$. This value, however, is too low for this process to be commercially acceptable. References [228-238] supply recent information on this topic. For the utilization of xylose in particular, see [18]. Soni [239] gives information about the use of *Cl. saccharoperbutylacetonicum*.

Production of single cell protein (SCP) is also well known: Schneider [240] fully describes production from pentoses. Lastly, the production of 2,3-butanediol by *Klebsiella pneumoniae* from pentoses could be of some interest and is described by Jansen and Tsao in this collection [17].

Possibly, the production of biopolymers can be realized using lignocellulosic hydrolyzates with *Alcaligenes eutrophus* and/or other microorganisms, but there is, at present, no information on the matter.

6 Ethanol Production Economics

The cost of ethanol production by the various acid processes discussed earlier is shown in Table 4. This table assumes that only the glucose is converted to ethanol. The most important observation is the similarity between the two dilute acid hydrolysis processes (plug flow and progressing batch) and the three concentrated acid hydrolysis processes (concentrated sulfuric, hydrochloric and hydrofluoric acid). The dilute acid processes have high feedstock costs because of the yield losses due to sugar degradation, but have low acid and base costs (because sulfuric acid is inexpensive, and only small amounts are used). Utility costs are low because the lignin and xylose, which are not utilized for ethanol production, are burned and provide a surplus of steam and electricity. Capital charges are high because a large and expensive processing plant is needed to convert the very large quantities of lignocellulosic (large quantities are necessary because of the low yield). Also important is the difference in selling price between the case in which furfural is sold as a chemical, and that in which furfural is burned at its fuel value to produce process heat. This points out the necessity of high process yields (fraction of the total feedstock converted to saleable product). Similar reductions in selling price would be achieved if the xylan fraction were reacted to xylose and converted to ethanol.

The concentrated acid processes also show similarities to each other. Because they convert virtually all the cellulose to utilizable sugars, their feedstock costs are lower.

Table 4. Data on production of ethanol from lignocellulosics

Hardwood consumption $t\,t^{-1}$	Process	Operating costs + Capital charges $ t^{-1}
7.6	Percolation	220
6.1	Progressing batch	181
8.2	Plug flow	204
8.2	id. (furfural sold)	33
5.3	H_2SO_4 conc.	330
5.4	HCl liquid	302
4.9	HF liquid	326
4.6	Enzymatic SHF (*)	450
3.2	Enzymatic SSF (*)	250

(*)-Hemicellulose used for ethanol production

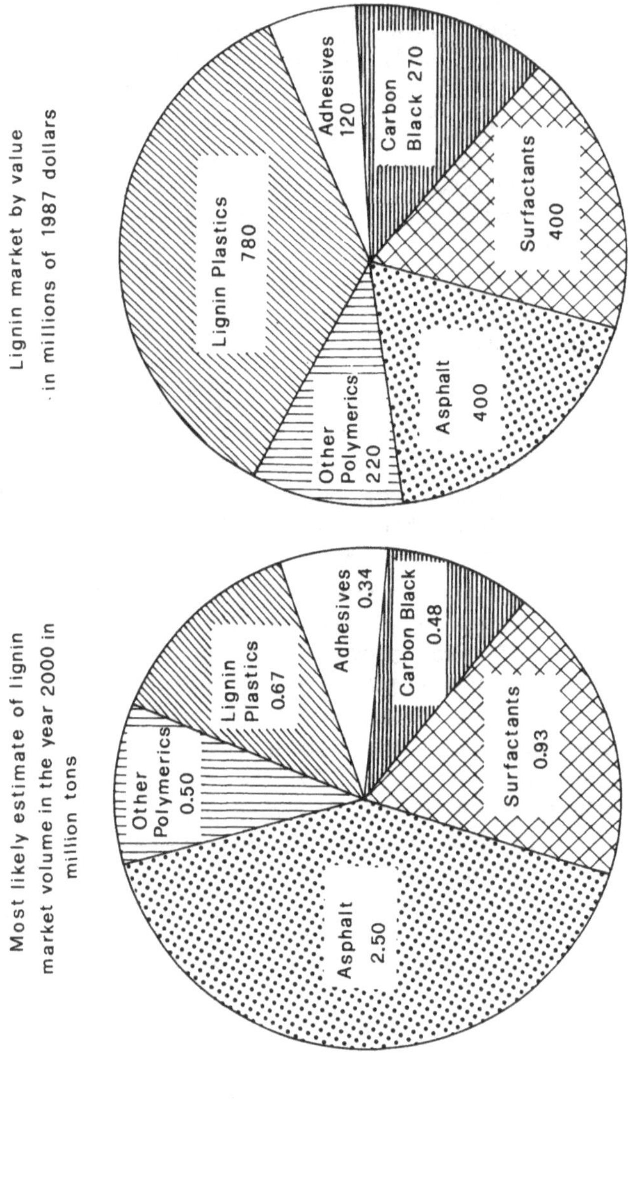

Fig. 12. Projected lignin market in the U.S. for the year 2000 [from [210]]

However, because they use expensive acids such as HCl or HF (some of which is inevitably lost due to neutralization by ash, or losses from the system) or because they use large amounts of an inexpensive acid (H_2SO_4), the costs for acid consumption are quite high. The capital costs are similar to those for dilute acid processes. Although the overall plant size for concentrated acid processes is smaller (higher yields mean less material has to be processed), the capital charges are still very large (due to the large expense involved in acid recovery or drying the biomass prior to the reaction).

The cost of ethanol production by these processes is quite high, to the order of \$ 0.43 to \$ 0.53 kg^{-1} (\$ 1.30 to 1.60 US $gallon^{-1}$) when xylose is not converted to a saleable product. Although some opportunities exist for improving the basic acid hydrolysis technology, these opportunities are limited because these processes have received years of development. The most important improvement which can be achieved in these systems is to introduce a xylose utilization process (e.g. for ethanol production). This will decrease the feedstock costs, dramatically reduce the processing plant size and capital investment, and decrease the amount of energy and labor needed to produce a given amount of ethanol. Further improvements would be possible if the lignin were also used beneficially (Fig. 12). Opportunities for this are greatest with the halogen acid processes (which do not greatly modify the lignin), somewhat less with the high temperature dilute acid process, and lowest with the progressing batch and concentrated sulfuric acid processes which extensively modify the lignin. For the economic analysis of xylose utilization see [251].

Cost of production summaries for the separate hydrolysis and bioconversion (SHF) and simultaneous saccharification and bioconversion (SSF) processes are shown in Figs. 13 and 14. In the SHF the major costs are feedstock, energy and capital charges, enzyme production, and the offsite systems for environmental control and utility (steam and electricity) production. The feedstock shown in near the lower limit, as it accounts for only the sugars actually converted to ethanol. A hidden feedstock cost is that of energy. Approximately \$ 0.20^{-1} kg (\$ 0.60 US $gallon^{-1}$) is converted into

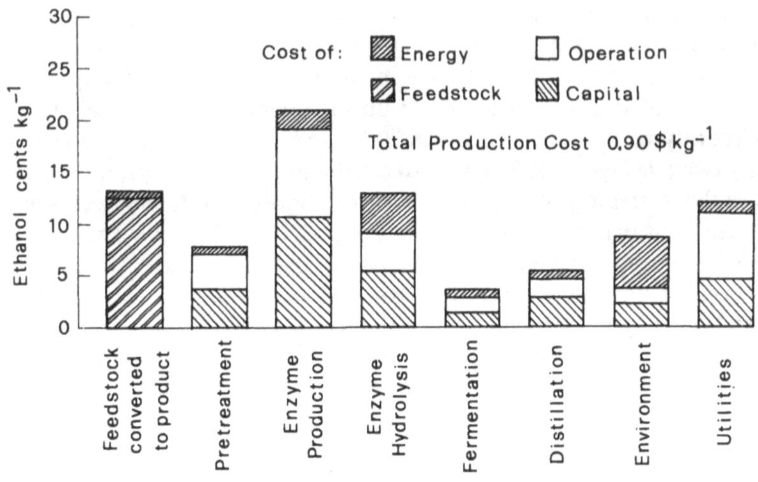

Fig. 13. Breakdown of ethanol production costs by process area for the SHF process [from [97]]

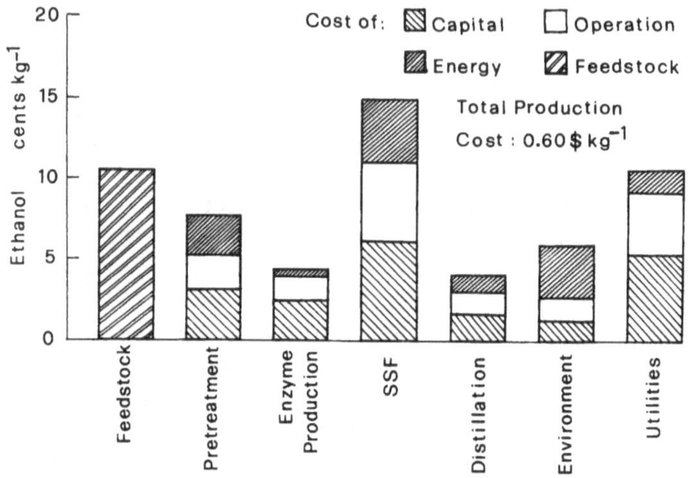

Fig. 14. Breakdown of ethanol production costs by process area for the SSF process [from [246]]

steam and electricity to run the process instead of into useful products. This is a consequence of the low process yield (only 70% of the cellulose, and none of the xylose or lignin are converted into saleable products with current technology). Capital charges are high, because a large processing plant is necessary to convert large volumes at low yields, and because of the large equipment sizes needed to process the dilute product streams coming from the hydrolysis reactor. Enzyme production costs are high because large amounts of enzyme are used to help overcome the effects of end product inhibition by cellobiose and glucose. Finally, the cost of cleaning up the waste streams and generating the steam and electricity is high because of the large volumes which must be processed (anything which is not converted into a saleable product must be processed either through the anaerobic digestion section or the boiler/steam generator).

The cost of production for SSF process is considerably lower. All cost reductions have their root in the reduction of end product inhibition due to the continuous removal of glucose and cellobiose by bioconversion. First, the yield is improved because the reaction is not stopped by the build up of sugars. Approximately 90% of the cellulose is hydrolyzed in a SSF process vs 70% in SHF. Also, product concentrations are roughly twice as high in SSF. This reduces the size, capital cost, and energy consumption of the downstream process by a factor of almost two. Energy costs are also reduced because considerably more ethanol is produced while the amount of material processed is reduced. The enzyme production costs are reduced by almost a factor of five, because enzyme loading can be reduced to one seventh of that necessary when inhibition by glucose must be overcome. Finally, because less of the initial feed must be processed in the waste treatment and boiler plants, the cost of these auxiliary sections are reduced. Taken together, these process improvements reduce the predicted price of ethanol from 0.90 to \$ 0.60 kg^{-1} (from 2.70 to \$ 1.80 US gallon^{-1}). Unlike acid hydrolysis, the technology of enzymatic hydrolysis is relatively new, and there is considerable room for improvement in each of the process areas. The interaction between the pretreatment and hydrolysis rate is still not completely understood. Better understanding of this relationship should allow higher hydrolysis rates to be achieved, as rates for the cellulose hydrolysis are still approximately two orders of magnitude below those of starch hydrolysis. Also, current pretreatment methods degrade up 50% of the C_5 fraction. This loss directly reduces the potential for ethanol production by xylose bioconversion and also produces degradation products which inhibit biotransformation.

Enzyme production shows considerable room for improving the process. Current designs use solid substrates, or substrates with limited supplies such as lactose. Research is needed to produce enzymes at high rates from lignocellulosic derived sugars such as glucose or xylose. A factor of three improvement in productivity should be achievable with suitable mutants. Similarly, cellulase is, as it was said, a mixture of three different types of individual enzymes and the total activity of the enzyme is dependent on the relative proportions of the three components in the mixture. It is not yet certain what proportions are correct for an enzyme mixture used in biomass processing, although it appears that mixtures with greater quantities of β-glucosidase are preferred. Optimization of these ratios should lower the total amount of enzyme which is needed, and improve the rate and yield of the reaction. The hydrolysis reaction can be improved by better matching the reaction conditions to those needed by the enzyme and yeast. Also the amount of energy needed to stir the reactor may be considerably less than is currently assumed. Continuous removal of the ethanol product should reduce inhibition of both the enzyme and the yeast, improving both rate and yield.

Finally, as in the case of acid hydrolysis, use of the other two major fractions (hemicellulose and lignin) would bring about the greatest reductions in ethanol cost. The cumulative effect of these improvements is shown in Fig. 15.

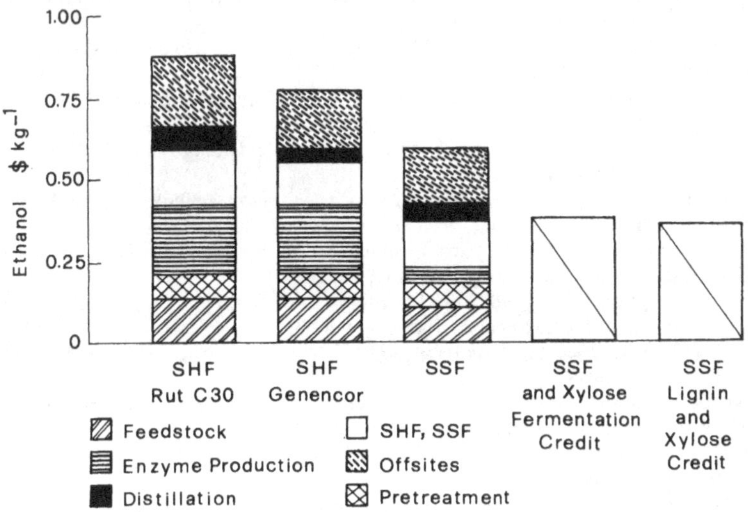

Fig. 15. Comparison of costs by process area for SHF and SSF process and cost reduction of SSF process if xylose and or lignin are sold [from [246)]]

7 Conclusions

Biological processes for the conversion of lignocellulosics to ethanol are attractive because of their potentially high efficiency. However, this means that the process must be configured to use all the major fractions of the feedstock: cellulose, hemicelluloses, lignin. The cellulose and hemicelluloses can be broken down to sugars by either acid or enzymatic hydrolysis processes. Acid processes are more developed, but enzymatic processes have the greater potential for improved performance. In the past decade, xylose has gone from being regarded as recalcitrant to ethanol to a point where 70% of the theoretical yield can be achieved at reasonable ethanol concentrations. Finally, methods have been identified for the conversion of lignins to liquid fuels.

Using advanced enzymatic hydrolysis processes in conjunction with xylose bioconversion, ethanol cost of approximately \$ 0.33 to 0.34 kg^{-1} (\$ 1.00 US $gallon^{-1}$) should be achievable. Addition of lignin processing steps should further reduce this cost.

This report could not have been prepared without the invaluable advice of John D. Wright, of the Solar Energy Research Institute, Golden, Colorado. His help is gratefully acknowledged.

8 References

1. Reese, ET, Mandels M, Weiss AH (1972) in: Ghose TK, Fiechter A (eds) Springer, Berlin Heidelberg New York, p 181 (Advances in biochemical engineering, vol 2)
2. Ghose TK, Das K (1972) in: Ghose TK, Fiechter A (eds) Springer, Berlin Heidelberg New York, p 55 (Advances in biochemical Engineering, vol 1)
3. Enari TM, Markkanen P (1977) in: Ghose TK, Fiechter A (eds) Springer, Berlin Heidelberg New York, p 1 (Advances in biochemical engineering, vol 5)
4. Linko M (1977) in: Ghose TK, Fiechter A (eds) Springer, Berlin Heidelberg New York, p 25 (Advances in biochemical engineering, vol 5)
5. Ghose TK (1977) in: Ghose TK, Fiechter A (eds) Springer, Berlin Heidelberg New York, p 39 (Advances in biochemical engineering, vol 6)
6. Fan LT, Lee Y-H, Beardmore DH (1980) in: Fiechter A (ed) Springer, Berlin Heidelberg New York, p 101 (Advances in biochemical engineering, vol 14)
7. Lee Y-H, Fan LT (1980) in: Fiechter A (ed) Springer, Berlin Heidelberg New York, p 101 (Advances in biochemical engineering, vol 17)
8. Lee Y-H, Fan LT, Fan LS (1980) in: Fiechter A (ed) Springer, Berlin Heidelberg New York, p 131 (Advances in biochemical Engineering, vol 17)
9. Chang MM, Chou TYC, Tsao GT (1981) in: Fiechter A (ed) Springer, Berlin Heidelberg New York, p 15 (Advances in biochemical engineering, vol 20)
10. Fan LT, Lee Y-H, Gharpuray MM (1982) in: Fiechter A (ed) Springer, Berlin Heidelberg New York, p 155 (Advances in biochemical engineering, vol 23)
11. Gong Ch-Sh et al. (1981) in: Fiechter A (ed) Springer, Berlin Heidelberg New York, p 93 (Advances in biochemical engineering, vol 20)
12. Kosaric N, Duvniak Z, Stewart GG (1981) in: Fiechter A (ed) Springer, Berlin Heidelberg New York, p 119 (Advances in biochemical engineering, vol 20)
13. Jeffries TW (1983) in: Fiechter A (ed) Springer, Berlin Heidelberg New York, p 1 (Advances in biochemical engineering/biotechnology, vol 27)
14. McCracken LD, Gong Ch-Sh (1983) in: Fiechter A (ed) Springer, Berlin Heidelberg New York, p 33 (Advances in biochemical engineering/biotechnology, vol 27)
15. Schneider H et al. (1983) in: Fiechter A (ed) Springer, Berlin Heidelberg New York, p 57 (Advances in biochemical engineering/biotechnology, vol 27)

16. Kurtzmann CP (1983) in: Fiechter A (ed) Springer, Berlin Heidelberg New York, p. 73 (Advances in biochemical engineering/biotechnology, vol 27)
17. Jansen NB, Tsao GT (1983) in: Fiechter A (ed) Springer, Berlin Heidelberg New York, p 85 (Advances in biochemical engineering/biotechnology, vol 27)
18. Volesky B, Szczesny T (1983) in: Fiechter A (ed) Springer, Berlin Heidelberg New York, p 101 (Advances in biochemical engineering/biotechnology, vol 27)
19. Janshekar H, Fiechter A (1983) in: Fiechter A (ed) Springer, Berlin Heidelberg New York, p 119 (Advances in biochemical engineering/biotechnology, vol 27)
20. Magee RJ, Kosaric N (1985) in: Fiechter A (ed) Springer, Berlin Heidelberg New York, p 61 (Advances in biochemical engineering/biotechnology, vol 32)
21. Franzidis JP, Porteous A (1981) in: Klass DL, Emert GH (eds) Fuels from biomass and wastes. Ann Arbor Science, Kent, p 267
22. Demuth R von (1913) Z. angew. Chem. 26: 786
23. Kressman FW (1922) US Dpt of Agriculture Bull. 983
24. Scholler H US Patents 1 641 771 (1927); 2 083 347 (1934); 2 083 348 (1934); 2 188 192 (1937); 2 188 193 (1937)
25. Scholler H Brevets Français 706 678 (1930); 799 358 (1936)
26. Scholler H (1936) Chem.-Ztg 60: 293
27. Scholler H (1939) DRP 676 967
28. Harris EE, Beglinger E (1946) Ind. Eng. Chem. 38: 890
29. Harris EE et al. (1945) Ind. Eng. Chem. 37: 12
30. Bergius F (1937) Ind. Eng. Chem. 29: 247
31. Fredenhangen K, Helferich B (1927) DRP 560 535 to IG Farbenindustrie
32. Helferich B, Böttger S (1929) Liebigs Ann. Chem. 476: 510
33. Saeman JF (1945) Ind. Eng. Chem. 37: 53
34. Root DF, Saeman JF, Harris JH (1959) Forest Products J. 9(5): 158
35. Fagan RD et al. (1971) Environm. Sc. Technol. 5: 545
36. Grethlein HE (1983) Acid hydrolysis of cellulosic biomass. A Progress Report. SERI Subcontract No XX-2-02-02152-1. NH: Dartmouth College, Hanover USA
37. Wright JD, d'Agincourt CG (1984) Evaluation of sulfuric acid hydrolysis processes for alcohol fuel production. SERI/TR-231-2074. Solar Energy Research Institute. Golden CO
38. Wright JD, Power AJ (1986) Comparative technical evaluation of acid hydrolysis processes for conversion of cellulose to alcohol. Presented at the Conference on Energy from Biomass, Institute of Gas Technology, Washington DC
39. Regnault A et al. (1978) DRP 2 814 067 to Battelle Memorial Institute
40. Foster AV, Martz LE, Leng DE (1980) US Patent 4 237 110 to Dow Chemical Co
41. Sklarewitz ML, Goldstein IS (1983) Recycle of hydrochloric acid in a wood hydrolysis plant by membrane technology. Presented at the Interamerican Congress of Chem. Eng., Santiago, Chile
42. Kusama J (1979) Chemical Econom. and Eng. Rev. 11(6): 32
43. Antonoplis RA, Blanch HW, Wilke CR (1983) High pressure HCl conversion of cellulose to glucose. LBL-1422 Lawrence Berkeley Labs, Berkeley CA. Available from NTIS, Springfield VA
44. Antonoplis RA et al. (1983) Biotechnol. Bioeng. 25: 2757
45. Defaye D, Gadelle G, Pedersen C (1981) Degradation of cellulose with hydrogen fluoride. In: Palz W, Chartier P, Hall DO (eds) Energy from biomass. Proceedings of the 1st ECC Conference London, Applied Science, London, p 292
46. Bentsen T (1983) Hydrolysis of carbohydrates in straw using hydrogen fluoride pretreatment. In: Strub A, Chartier P, Schleser J (eds) Energy from biomass. Proceedings of the 2nd ECC Conference Berlin, p 1103. Applied Science, London
47. Downey K et al (1983) HF hydrolysis of wood for ethanol production. Presented at Industrial Energy Forum 83, September 19, Nashville TN
48. Fredenhangen K, Cadenbach G (1933) Angew. Chem. 46: 113
48. Selke SM, Hawley M, Lamport DTA (1983) Wood and Agricultural Residues 329
50. Franz R et al. (1983) Lignocellulose saccharification by HF. In: Strub A, Chartier P, Schleser J (eds) Energy from biomass. Proceedings of the 2nd ECC Conference, Berlin, Applied Science, London, p 873

51. Ostrovski CM, Aitken J, Free D (1984) New developments in fuel ethanol production by gaseous anhydrous hydrofluoric acid hydrolysis of hardwood. Presented at Bioenergy 84, Gathenburg Sweden
52. Dunning JW, Lathrop EC (1945) Ind. Eng. Chem. 37: 24
53. Dunning JW, Lathrop EC (1948) US Patent 2 450 586
54. US Dpt of Interior (1951) Liquid fuels from agricultural residues. Rpt of Investigations 4772-Part III. US Dpt of Interior, Washington DC
55. Wenzl HFJ (1970) in: The chemical technology of wood, Academic New York, p. 32
56. Ackerson M, Ziobro M, Gaddy GL (1981) Biotechnol. Bioeng. Symp. 11: 103
57. Tsao GT et al. (1982) Process Biochem. (IX-X): 34
58. Wright JD (1983) Design and evaluation of low-temperature, concentrated acid hydrolysis process. SERI/TR-231-1913. Solar Energy Research Institute, Golden, CO
59. Barrier JW et al. (w.d.) Experimental production of ethanol from agricultural cellulosic materials using low temperature acid hydrolysis. Tennessee Valley Authority, Muscle Shoals AL
60. Badger PC et al. (1984) in: Preprints of the 6th International Symposium on Alcohol Fuel Technology, Ottawa, Canada, vol 2 p 100
61. Farina GE, Barrier JW, Forsythe ML (1986) in: Proceedings of the 7th International Symposium on Alcohol Fuel Technology, Technip, Paris, p 44
62. Brenner W et al. (1979) Radiat. Phys. Chem. 14: 299
63. Thompson DR, Grethlein HE (1979) Ind. Eng. Chem. Prod. Res. Dev. 18: 166
64. Church JA, Wooldridge D (1981) Ind. Eng. Chem. Prod. Res. Dev. 20: 378
65. Teng KF, Mutharasan R, Grossmann ED (1983) Paper presented at the AIChE Annual Meeting Oct. 30-Nov. 3, Washington DC
66. Grethlein HE (1978) Biotechnol. Bioeng. 20: 503
67. Sharman DK (1984) Two-step process for the selective production of fermentable sugars and ethanol from biomass residues. In: Prepr. of the 6th International Symposium on Alcohol Fuel Technology, Ottawa, Canada, vol. 2 p 205
68. Horwath J, Mutharasan R, Grossmann ED (1983) Biotechnol. Bioeng. 25: 19
69. Wright JD (1983) High temperature acid hydrolysis of cellulose for alcohol production. SERI/TP-231-2058. Solar Energy Research Institute, Golden CO
70. Badger Engineers Inc (1984) Economic feasibility study of an acid based ethanol plant. SERI Subcontract No ZX-3-030-96-2. Badger Eng. Inc., 1 Broadway, Cambridge MA
71. Ritter FJ (1984) US Patent 427453
72. Singh A, Das K, Sharma DK (1984) Ind. Eng. Chem. Prod. Res. Dev. 23: 257
73. Harris JF et al. (1985) Two-stage sulfuric acid hydrolysis of wood. USDA Forest Products Labs, Madison WI
74. Hokanson AE, Katzen R (1978) Chem. Eng. Progr. 74: 67
75. Mendelsohn HR, Wettstein P (1981) Chem. Eng. 62
76. Burton RJ (1982) The New Zealand wood hydrolysis process. In: Preprints of the Ethanol from Biomass Conference, Winnipeg Canada
77. Mackie K, Deverell K, Callander I (1982) Aspects of wood hydrolysis via the dilute sulfuric acid process. Preprints of the Ethanol from Biomass Conference, Winnipeg Canada
78. Uprichard JM, Burton RJ (1982) Ethanol from wood. Preprints of the 5th International Symposium on Alcohol Fuel Technology, Auckland New Zealand, p 317
79. Carvalho Neto CC, Kling SH (1983) Pilot plant acid studies using *Eucalyptus paniculata*. Presented at the Interamerican Congress of Chem. Eng., Santiago Chile
80. Beck MJ, Strickland RC (1984) Biomass 6: 101
81. Raymond B (1986) in: Proceedings of the 7th International Symposium on Alcohol Fuel Technology, Technip, Paris, p 39
82. Wright JD, Bergeron PW, Wendene PJ (1985) The progressing batch hydrolysis reactor. SERI/TP-232-2803. Solar Energy Research Institute, Golden CO
83. Bergeron PW, Wright JD, Werdene PJ (1986) Biotechnol. Bioeng. Symp. 17: 33
84. Leonard RH, Hainj GJ (1945) Ind. Eng. Chem. 37: 390
85. Lüers H et al. (1937) Z. Spiritusind. 60: 7
86. Strickland RC, Beck MJ (1985) Effective pretreatments and neutralization methods for ethanol production from acid-catalyzed hardwood hydrolysates using *Pachysolen tannophilus*. Presented at the 9th Symposium on Energy from Biomass and Wood Wastes, Lake Buena Vista FL

87. Converti A et al. (1985) Biotechnol. Bioeng. 27: 1108
88. Del Borghi M et al (1985) Biotechnol. Bioeng. 27: 761
89. Reese ET, Sin RGH, Levinson HS (1950) J. Bacteriol. 59: 485
90. Bruinenberg PM et al. (1983) J. Appl. Microbiol. Biotechnol. 18: 278
91. Poutanen K, Pouls J, Linko M (1987) in: Proceedings of the 4th European Congress on Biotechnology, Elsevier, Amsterdam, vol 3 p 90
92. Himmel M (1986) Biotechnol. Bioeng. Symp. 17: 413
93. Bailey MJ, Nevalainen KMH (1981) Enz. Microbiol. Technol. 3: 153
94. Ghose TK, Ghosh P (1978) J. Appl. Chem. Biotechnol. 28: 309
95. Woodward J, Wiseman A (1982) Enz. Microbiol. Technol. 4: 73
96. Alfani F et al. (1983) Biodegradation of native cellulose. In: Palz W, Coombs J, Hall DO (eds) Energy from biomass. Proceedings of the 3rd ECC Conference Venice, Elsevier, Amsterdam, p 989
97. Wright JD, Power AJ, Douglas LJ (1986) Biotechnol. Bioeng. Symp. 17: 285
98. Takagi M et al. (1977) A method for production of alcohol directly from cellulose using cellulase and yeast. In: Proceedings of the Bioconversion Symposium Delhi, ITT, New Delhi, p 551
99. Rivers DB, Emert GH (1980) in: Preprints of the Bioenergy 80 World Congress and Exposition Atlanta GA, p 157
100. Pemberton MS, Brown RD Jr, Emert GH (1980) Can. J. Chem. Eng. 58: 273
101. Becker DK, Blotkamp PJ, Emert GH (1981) Pilot scale conversion of cellulose to ethanol. In: Klass DL, Emert GH (eds) Ann Arbor Science, Ann Arbor MI, p 375 (Fuels from Biomass and Wastes)
102. Gonde P et al. (1984) Appl. Environm. Microbiol. 48: 265
103. Lastick SM (1984) Simultaneous saccharification and fermentation of cellulose. In: Biotech 84, p 277
104. Ghosh P (1984) Biotechnol. Bioeng. 26: 377
105. Spangler DJ, Emert GH (1986) Biotechnol. Bioeng. 28: 115
106. Favela Torres E, Baratti JC (1987) Appl. Microbiol. Biotechnol. 27: 121
107. Gonde P et al. (1984) in: Proceedings of the 3rd European Congress on Biotechnology, Verlag Chemie, Weinheim, vol 2 p 15
108. Wyman CE et al. (1986) Biotechnol. Bioeng. Symp. 17: 221
109. Spindler DD et al. (1987) Thermotolerant yeast for simultaneous saccharification and fermentation of cellulose to ethanol. Presented at the 9th Symposium on Biotechnology for Fuels and Chemicals, Boulder Colorado. To be published in Biotechnol. Bioeng. Symp.
110. Dekker RFH, Wallis AF (1983) Biotechnol. Bioeng. 25: 3027
111. Foody E (w. d.) Scale-up testing of Iotech's Enzyme Production Process using chemical grade feedstock and the RL-P37 Organism. Contractor's Final Report to Energy, Mines and Resources, Canada, Contract No 0250.23216-3-6264
112. Pourquié J (1987) Lignocellulosic biomass valorisation. Scale-up of the enzymatic process. In: Grassi G et al. (eds) Biomass for energy and industry. Proceedings of the 4th ECC Conference Orléans. Elsevier, London, p 341
113. Fähnrich P, Irrgang K (1981) Biotechnol. Lett. 3: 201
114. Fähnrich P, Irrgang K (1982) Biotechnol. Lett. 4: 519, 775
115. Leisola MSA et al. (1985) Biotechnol. Bioeng. 27: 1389
116. Eriksson KE, Petterson B (1971) Biodeterioration of materials. Applied Science, London, vol 2 p 116
117. Wood TM (1969) Biochem. J. 113: 457
118. Wood TM, McCrae SI (1977) Carbohydr. Res. 57: 117
119. Madan N, Sood P (1979) Microbiol. Lett. 12: 109
120. Targonski Z, Szajer C (1979) Biotechnol. Lett. 1: 75, 439
121. White WL et al. (1948) Mycologia 40: 34
122. Veng PP, Gong CS (1982) Enz. Microbiol. Technol. 4: 169
123. Macris BJ, Kekos K, Evangelidou X (1987) Enhanced cellulase activity of *Fusarium oxysporum* grown on straw for ethanol production. In: Grassi G et al. (eds) Biomass for energy and industry. Proceedings of the 4th ECC Conference Orléans, Elsevier, London, p 699
124. Wood TM, Wilson CA (1987) The rumen of a sheep: a new source of cellulase for producing

fermentable glucose from cellulosic wastes. In: Grassi G et al. (eds) Biomass for energy and industry. Proceedings of the 4th ECC Conference Orléans, Elsevier, London, p 727

125. Salby K, Maitland GC (1971) Biochem. J. 104: 716
126. Wood TM et al. (1987) Maximizing the production of fermentable soluble sugars from straw using enzymes synthesized by the Fungus *Penicillium pinophilum*. In: G. Grassi et al. (eds) Biomass for energy and industry. Proceedings of the 4th ECC Conference Orléans, Elsevier, London, p 732
127. Wood TM, Brown JA, McCrae SI (1987) in: Proceedings of the 4th European Congress on Biotechnology, Elsevier, Amsterdam, vol 2 p 335
128. Chahal DS, Hawksworth DL (1976) Mycologia 68: 600
129. McHale A, Coughlan MP (1980) FEBS Lett. 117: 318
130. Jain S, Tiraby G (1987) Separation and characterisatization of the cellulolytic components of a thermophilic Fungus: *Talaromyces* sp. CL-240. In: G. Grassi et al. (eds) Biomass for energy and industry. Proceedings of the 4th ECC Conference Orléans, Elsevier, London, p 355
131. McHale AP, Morrison J, McCarthy U (1987) Production of cellulase by *Talaromyces emersonii* CBS 814.70 and a mutant UV7 during growth on lactose. In: G. Grassi et al. (eds) Biomass for energy and industry. Proceedings of the 4th ECC Conference Orléans, Elsevier, London, p 704
132. Durand H, Soucaille P, Tiraby G (1984) Enz. Microbiol. Technol. 6: 175
133. Saddler JN, Khan AW (1980) Can. J. Microbiol. 26: 760
134. Armstrong DW, Brown DA, Martin GM (1982) in: Proceedings of the 4th. Bioenergy Research and Development Seminar. Winnipeg, Canada
135. Cooney CL et al. (1978) Biotechnol. Bioeng. Symp. 8: 103
136. Brooks R et al. (1979) in: 3rd Annual Biomass Energy System Proceedings. SERI, Golden CO
137. Shinmyo A, Garcia-Martinez V, Demain AL (1979) J. Appl. Biochem. 1: 202
138. Ng TK, Ben Bassat A, Zeikus JG (1981) Appl. Environm. Microbiol. 41: 1337
139. Schwarz WH et al. (1987) Appl. Microbiol. Biotechnol. 27: 50
140. Holtzapple M, Humphrey AE, Sye EK (1980) in: Preprints of the 6th International Fermentation Symposium, London, Canada, p 87
141. Phillips JA, Humphrey AE (1980) in: Preprints of the 2nd International Symposium on Bioconversion and Biochemical Engineering Delhi. ITT, New Delhi
142. Malfait M, Godden B, Pennincks M (1984) Ann. Microbiol. 135B: 79
143. McCarthy AJ, Ball AS (1987) Lignocellulose degradation by *Actynomycetes*. In: Grassi G et al. (eds) Biomass for energy and industry. Proceedings of the 4th ECC Conference Orléans, Elsevier, London, p 351
144. Halliwell G, Phillips T (1987) Synthesis and activity of cellulase produced by strains of the thermophilic, anaerobic, cellulolytic bacterium *Clostridium thermocellum*. In: Grassi G et al. (eds) Biomass for energy and industry. Proceedings of the 4th ECC Conference Orléans, Elsevier, London, p 709
145. Béguin P et al. (1987) Conversion of cellulose into ethanol by *Clostridium thermocellum*: Genetic engineering cellulase. In: G. Grassi et al. (eds) Biomass for energy and industry. Proceedings of the 4th ECC Conference Orléans, Elsevier, London, p 346
146. Faure E et al. (1987) Caractérisation du système cellulolytique du *Clostridium thermocellum*. In: G. Grassi et al. (eds) Biomass for energy and industry. Proceedings of the 4th ECC Conference Orléans, Elsevier, London, p 717
147. Blotkamp PJ et al. (1978) AIChE Symp. Series 74 (181): 85
148. Downing KM, Ho CH, Zabriskie DW (1987) Biotechnol. Bioeng. 29: 1086
149. Cowling EG, Kirk TK (1976) Biotechnol. Bioeng. Symp. 6: 59
150. Hendy NW, Wilke CR, Blanch H (1984) Enz. Microbiol. Technol. 6: 73
151. Taylor JD (1981) in: Palz W, Chartier P, Hall DO (eds) Energy from biomass. Proceedings of the 1st ECC Conference London. Applied Science, London, p 330
152. Sinner M, Schreier M, Ballweg A (1983) in: Strub A, Chartier P, Schleser J (eds) Energy from biomass. Proceedings of the 2nd ECC Conference Berlin, Applied Science, London, p 984
153. Wayman M (1980) in: Proceedings of the 4th International Symposium on Alcohol Fuel Technology, Guarujà, Brazil, vol 1 p 79
154. Delong EA (1981) Can. Patent 1 096 374

155. Foody E (w.d.) Optimization of steam explosion pretreatment. Final Report to US Department of Energy (Contract No DE-AC-02-79-ET 23050)
156. Pourquié J, Glinkmans G (1986) in: Proceedings of the 7th International Symposium on Alcohol Fuel Technology, Technip, Paris, p 54
157. Grohmann K, Torget R, Himmel M (1985) Biotechnol. Bioeng. Symp. 15: 59
158. Grethlein HE, Allen DC, Converse AO (1984) Biotechnol. Bioeng. 26: 1498
159. Torget R et al. (1988) Initial design and parametric evaluation of a dilute acid pretreatment process for aspen wood chips. Appl. Biochem. Biotechnol. (in press)
160. Brownell HH, Yu EKC, Saddler JN (1986) Biotechnol. Bioeng. 28: 792
161. Sarkanen KV (1980) in: Acid catalyzed delignification of lignocellulosics in organic solvents, Academic, New York, vol 2 p 127
162. Cunningham RL, Carr ME, Bagby MO (1985) Biotechnol. Bioeng. Symp. 15: 17
163. Leisola MSA, Fiechter A (1985) Adv. Biotechnol. Processes 5: 59
164. Sinitsyn AP, Bougay HR, Clesceir LS (1983) Biotechnol. Bioeng. 25: 1393
165. Chahal DS (1986) in: Proceedings of the 7th International Symposium on Alcohol Fuel Technology, Technip, Paris, p 39
166. Tjerneld et al. (1985) Biotechnol. Bioeng. Symp. 15: 419
167. Tan LUL et al. (1986) Appl. Microbiol. Biotechnol. 25: 250, 256
168. Cantarella M et al. (1979) Biochem. J. 179: 15
169. Alfani et al. (1986) in: Magnien E (ed) Biomolecular engineering in the European Community, Martinus Hjihoff, Dordrecht, Netherland, p 143
170. Spasov SD (1987) Efficient hydrolysis of natural lignocellulosics materials by a *Trichoderma viride* cellulase immobilized on various soluble polymers. In: Proceedings of the 4th European Congress on Biotechnology, Elsevier, Amsterdam, vol 2 p 101
171. Cantarella M et al. (1987) A comparative study on cellulase immobilization in synthetic and natural gels. In: Proceedings of the 4th European Congress on Biotechnology, Elsevier, Amsterdam, vol 2 p 112
172. Boidin J, Adzet JM (1957) Bull. Soc. Mycol. France 73: 331
173. Schneider H (1981) Biotechnol. Lett. 3: 89
174. Slininger PJ et al. (1982) Biotechnol. Bioeng. 24: 371
175. Detroy RW et al. (1982) Biotechnol. Bioeng. 24: 1105
176. Dekker RFH et al. (1982) Biotechnol. Lett. 4: 411
177. Chung IS, Lee Y-Y, Beck MJ (1986) Biotechnol. Bioeng. Symp. 17: 391
178. Alexander NJ (1986) Appl. Microbiol. Biotechnol. 25: 203
179. du Preez JC, van der Waalt JP (1983) Biotechnol. Lett. 5: 537
180. du Preez JC, Prior BA, Monteiro MT (1984) Appl. Microbiol. Biotechnol. 19: 261
181. Toivola A et al. (1984) Appl. Environm. Microbiol. 47: 221
182. Jeffries TW (1985) Biotechnol. Bioeng. Symp. 15: 149
183. Wayman M, Tsuyuki ST (1985) Biotechnol. Bioeng. Symp. 15: 167
184. Wayman M et al. (1986) in: Proceedings of the 7th International Symposium on Alcohol Fuel Technology, Technip, Paris, p 597
185. Lucas C, van Uden N (1985) J. Basic Microbiol. 25: 547
186. van Zyl C, Prior BA, du Preez JC (1987) Production of ethanol from sugar cane bagasse hemicellulose hydrolysate by *Pichia stipitis*. Presented at the 9th Symposium on Biotechnology for Fuels and Chemicals, Boulder CO. To be published in Biotechnol. Bioeng. Symp.
187. Rosnick CS, Chung IS, Lee Y-Y (1987) High-cell continuous fermentation of xylose to ethanol by *Pichia stipitis* using cell-recycled bioreactor. Presented at the 9th Symposium on Biotechnology for Fuels and Chemicals, Boulder Colorado. To be published in Biotechnol. Bioeng. Symp.
188. Alexander MA, Jeffries TW (1987) Effect of ethanol on continuous fermentation of xylose by *Candida shehatae*. Presented at the 9th Symposium on Biotechnology for Fuels and Chemicals, Boulder Colorado. To be published in Biotechnol. Bioeng. Symp.
189. Grba S, Besic S, Ban SN (1987). In: Proceedings of the 4th European Congress on Biotechnology. Elsevier, Amsterdam, vol 3 p 393
190. Dellweg H et al. (1984) Biotechnol. Lett. 6: 395
191. Rizzi M et al. (1987) in: Proceedings of the 4th European Congress on Biotechnology, Elsevier, Amsterdam, vol 3 p 415

192. Torrie JP, Wilson JJ (1987) Evaluation of fermenting yeasts and fungi using sugar substrates. Presented at the 9th Symposium on Biotechnology for Fuels and Chemicals, Boulder, Colorado. To be published in Biotechnol. Bioeng. Symp.
193. Snikho ML, Enari TM (1981) Biotechnol. Lett. 3: 273
194. Snikho ML, Suomalainen I, Enari TM (1983) Biotechnol. Lett. 5: 525
195. Tsao GT et al. (1981) Biotechnol. Bioeng. Symp. 11: 315
196. Ho NWY et al. (1983) Biotechnol. Bioeng. Symp. 13: 245
197. Huang JJ, Ho NWY (1983) Biochem. Biophys. Res. Comm. 126: 1154
198. Ho NWY (1985) Improvement of yeast xylose fermentation and utilization via genetic engineering. In: Biochemical Conversion Program Semi-Annual Review Meeting, SERI/CP-231-2726 Solar Energy Research Institute, Golden CO, p 295
199. Ueng PP et al. (1985) Biotechnol. Lett. 7: 153
200. Lastick SM et al. (1986) Biotechnol. Lett. 8: 1
201. Jeyaseelan K, Singh P (1987) in: Proceedings of the 4th European Congress on Biotechnology, Elsevier, Amsterdam, vol 1 p 394
202. Lastick SM et al. (1985) Xylose fermentation project. In: Biochemical Conversion Program Semi Annual Review Meeting, SERI/CP-231-2726. Solar Energy Research Institute, Golden CO, p 243
203. Hasche RL (1945) Chem. Eng. News 23 (20): 1840
204. Kirschenbaum I (1978) Butadiene. In: Kirk-Othmer Encyclopedia of Chemical Technology. Wiley, New York, vol 4 p 313
205. Delmas M, Gaset A (1986) Chimie et utilisation industrielle des pentoses et dérivés. Presented at the Symposium International sur l'Utilisation non Alimentaire du Blé et du Maïs, APRIA, Paris
206. BIOS (British Intelligence Objective Subcommittee) (1948) No 351
207. Reppe W (1953) Liebigs Ann. Chem. 582: 87
208. Mile NA, Walsh WL (1935) J. Am. Chem. Soc. 57: 1389
209. Parisi F (1987) Production of ethanol from biomass and its socio-economic impact. Presented at the 4th European Congress on Biotechnology, Amsterdam
210. Chum HL et al. (1985) The economic contribution of lignins to ethanol production from biomass. SERI/TR-231-2488. Solar Energy Research Institute, Golden CO
211. Parkhurst HJ, Huibers DTA Jr., Jones MW (1980) in: ACS Symp. Series, Div. Petroleum Chemistry, American Chemical Society, Washington DC, p 657
212. Gendler GL, Huibers DTA Jr, Parkhurst HJ (1983) in: Soltes E (ed) Wood and agriculture residues: Research on use for feed, fuels and chemicals, Academic, New York, p 391
213. Coughlin RW et al. (1984) In: Wise DC (ed) Bioconversion) systems, CRC Press, Boca Raton FL, p 41
214. Goheen DW (1971) in: Sarkanen KV, Ludwig CH (eds) Lignins: occurrence, formation, structure and reactions. Wiley, New York, p 797
215. Singeman GM (1980) Methyl aryl ethers from coal liquids as gasoline extenders and octane improvers. Gulf Research and Development Co, Pittsburgh PA
216. Hsu OH-H, Glasser WG (1975) Appl. Polymer Symp. 28: 297
217. Hsu OH-H, Glasser WG (1976) Wood Science 9(2): 97
218. Saraf VP, Glasser WG (1984) J. Appl. Polymer Sci. 29: 1831
219. Dilling P, Sargent P (1984) US Patent 4 454 066
220. Nimz HH (1983) in: Pizzi A (ed) Wood adhesives, Dekker, New York, p 248
221. Gillespie RH (ed) (1984) Adhesives for wood. Noyes Publications, Park Ridge NJ
222. Muller PC, Glasser WG (1984) J. of Adhesion 17: 157
223. Muller PC, Kelley SS, Glasser WG (1984) J. of Adhesion 17: 185
224. Glasser WG, Wu LC, Selin JF (1983) in: Soltes E (ed) Wood and agriculture residues: Research on use for feed, fuels and chemicals Academic, New York, p 149
225. Sirianni AF, Puddington IE (1972) Rubber World 165(6): 40
226. Dimitri MS (1976) US Patent 3 991 022
227. Sirianni AF, Puddington IE (1976) US Patent 3 984 362
228. Weizmann C (1912) British Patent 4 845
229. Andersch W, Bahl A, Gottschalk G (1982) in: Proceedings 5th Symposium Technische Mikrobiologie Berlin, Institut für Gärungsgewerbe und Biotechnologie, Berlin FRG, p 177

230. Vandecasteele JP, Pourquié J (1984) in: Preprints of the 6th International Symposium on Alcohol Fuel Technology, Ottawa, Canada, vol 2 p 227
231. Wayman M, Husted GR, Santangelo JD (1984) in: Preprints of the 6th International Symposium on Alcohol Fuel Technology, Ottawa, Canada, vol 2 p 234
232. Votruba J, Voleski B (1984) in: Proceedings of the 3rd European Congress on Biotechnology, Verlag Chemie, Weinheim FRG, vol 2 p 301
233. Larsson M, Holst O, Mattiasson B (1984) in: Proceedings of the 3rd European Congress on Biotechnology, Verlag Chemie, Weinheim FRG, vol 2 p 313
234. Dadgar AM, Foutch GL (1985) Biotechnol. Bioeng. Symp. 15: 611
235. Vandecasteele JP et al. (1986) in: Proceedings of the 7th International Symposium on Alcohol Fuel Technology, Technip, Paris, p 24
236. Pierrot P et al. (1987) in: Proceedings of the 4th European Congress on Biotechnology, Elsevier, Amsterdam, vol 1 p 258
237. Soni BK, Soucaille P, Goma G (1987) in: Proceedings of the 4th European Congress on Biotechnology, Elsevier, Amsterdam, vol 3 p 335
238. Fick M, Engasser GM (1987) in: Proceedings of the 4th European Congress on Biotechnology, Elsevier, Amsterdam, vol 3 p 509
239. Soni BK et al. (1986) Biotechnol. Bioeng. Symp. 17: 591
240. Schneider H et al. (1981) Pentose fermentation by yeasts. In: Stewart GG, Rennel I (eds) Current developments in yeast research. Advances in Biotechnology, Pergamon, Toronto, p 81
241. Chum HL et al. (1985) Evaluation of pretreatment of biomass for enzymatic hydrolysis of cellulose. SERI/TR-231-2183. Solar Energy Research Institute, Golden CO
242. Wright JD, Power AJ (1985) Biotechnol. Bioeng. Symp. 15: 511
243. Parker S et al. (1983) The value of furfural/ethanol coproduction from acid hydrolysis processes. SERI/TR-231-2000. Solar Energy Research Institute, Golden CO
244. Kosaric N et al. (1983) Ethanol fermentation. In: Dellweg H (ed) Biotechnology, Verlag Chemie, Weinheim FRG, vol 3 p 309
245. Isaacs SH (1984) Ethanol production by enzymatic hydrolysis. SERI/TR-231-2093. Solar Energy Research Institute, Golden CO
246. Wright JD, Wyman CE, Grohman K (1987) Simultaneous saccharification and fermentation of lignocellulose: Process evaluation. Presented at the 9th Symposium on Biotechnology for Fuels and Chemicals, Boulder CO. To be published in Biotechnol. Bioeng. Symp.

References added in proofs:

247. Brownbell HH, Saddler JN (1987) Biotechnol. Bioeng. 29: 228
248. Wong KY et al. (1988) Biotechnol. Bioeng. 30: 447
249. Wasson L et al. (1988) The effect of time and temperature on rapid steam hydrolysis (RASH). Presented at the 10th Symposium on Biotechnology for Fuels and Chemicals, Gatlinburg TN. To be published in Biotechnol. Bioeng. Symp.
250. Schultz TP, Rughani J, McGinnis JD (1988) A comparison of the pretreatment of sweetgum and white oak by steam explosion and RASH processes. Presented at the 10th Symposium on Biotechnology for Fuels and Chemicals, Gatlinburg TN. To be published in Biotechnol. Bioeng. Symp.
251. Hinman ND et al. (1988) Xylose fermentation: an economic analysis. Presented at the 10th Symposium on Biotechnology for Fuels and Chemicals, Gatlinburg TN. To be published in Biotechnol. Bioeng. Symp.

Modelling, Identification and Control of the Activated Sludge Process

Stefano Marsili-Libelli
Department of Systems and Computers Engineering, University of Florence, Via S. Marta, 3-50139 Firenze, Italy

1 Introduction ... 90
2 Reduced-Order Dynamic Modelling ... 91
2.1 Simplified Microbial Kinetics ... 92
2.2 Dissolved Oxygen Kinetics ... 94
2.3 Simplified Nitrification Kinetics ... 95
2.4 Sedimentation Dynamics ... 96
2.5 Continuous-Flow Model ... 101
3 Model Identification ... 103
3.1 SML Model Identifiability ... 103
3.2 Parametric Sensitivity ... 105
3.3 Practical Model Identification ... 108
3.4 On-line Parameter Identification ... 114
3.5 On-line Estimation of Bioactivities ... 117
3.6 On-line Estimation of Process Variables ... 120
4 Process Control ... 122
4.1 Process Performance Indicators ... 122
4.2 Conventional Control Strategies ... 123
4.2.1 PID Control ... 123
4.2.2 Approximate Optimal Control ... 124
4.3 Activated Sludge Control ... 127
4.3.1 Dissolved Oxygen Control ... 127
4.3.2 Sludge Recycle Control ... 131
4.3.3 SCOUR Control ... 137
4.4 Adaptive Control ... 138
4.4.1 Structure of the Dissolved Oxygen Adaptive Controllers ... 138
4.4.2 Performance of Self-tuning Controllers ... 140
5 Conclusions ... 143
6 List of Symbols ... 144
7 References ... 146

Practical experience shows that the efficiency and reliability of the activated sludge wastewater treatment process can be significantly improved if its time-varying nature is taken into account in the design of automatic control systems. This paper focuses on the dynamic aspects of the activated sludge treatment process and assesses the operational improvements brought about by automatic control. The three aspects of *modelling*, *identification* and *control* are reviewed and their interdependence is stressed. After introducing a simplified model for the pollutant/biomass interaction, the estimation and control problems are addressed and practical algorithms are discussed. The improvements brought about by such algorithms are clearly demonstrated.

1 Introduction

Biological wastewater treatment by means of activated sludge processes has been an expanding business for years and this may convey the notion of a well established engineering practice. Its widespread use might imply that such a technical expertise has been reached that any further significant advancement is unlikely to occur and that plant operation is exactly predictable and satisfactory.

A closer look at scores of operating records, though, reveals a dramatically different picture. Traditionally, sanitary engineers have put a premium on plant design rather than operation, and there are now countless reports of gross process failures caused by inadequate operational management. This, together with increasing energy costs and stricter environmental standards, has stressed the need for reliable automatic control.

In fact, the day-to-day operational practice has shown that relying on the steady-state design assumptions could be dangerously misleading given the inherently uncertain process dynamics and the large, unmeasurable variations of operating conditions. Regrettably, several factors have prevented the application of computer control to activated sludge treatment plants, compared with other biotechnological processes. The difficulty in obtaining reliable process measurements ranks highest among them, and the ensuing uncertainty in assessing the plant efficiency has surely discouraged any endeavour to improve it.

These considerations demonstrate that a deep division exists between design practice and process performance, but also indicates that the high operational standards now demanded can only be attained by combining sanitary engineering skills with a thorough knowledge of process dynamics and the application of advanced control strategies.

Since the pioneering research of Andrews [1-6] and Olsson [7,8] the scientific interest in the dynamic aspects of wastewater treatment processes has increased and several initiatives took place to assess the research needs in this area. References [9-11] collect papers presented at specialized scientific meetings to assess the impact and potential benefits of taking into account the dynamic nature of the problem in a comprehensive time-varying water quality management scheme.

Mathematical modelling, identification and real-time control of this biotechnological process represent a challenging area of endeavour for biotechnologists and control engineers alike. In fact, nonlinear process kinetics, time-varying parameters, and the lack of directly measurable process variables, all call for new and imaginative engineering solutions.

This paper reviews the research developments over the last decade concerning the three aspects of modelling, identification, and control of the activated sludge wastewater treatment process demonstrating how these aspects are tightly interconnected. Though mathematical modelling of microbiological systems is now a well established discipline (see e.g. [12,13]) and many descriptive mathematical models are available, few of these results can be applied to process control because of their complexity. As an alternative, this paper advocates the use of simplified models, which can be easily included in estimation and control algorithms and implemented on small, dedicated microcomputers.

The scope of the paper is evenly distributed among three main sections, each devoted with equal emphasis to modelling, identification and control of the process to stress their interdependence. In Sect. 2, a simplified model, developed by the author [14,15] is introduced. It is shown that this model compares well with the widely used Monod [16] kinetics, and can explain satisfactorily the interaction between pollutant and the specialized biomass. Section 3 describes the identification of this model, both theoretically and practically, with reference to numerical and experimental problems; the on-line estimation of relevant process quantities is also considered and practical estimation algorithms are presented and discussed. Section 4 focuses on the design of control laws for this specific process and reviews several possible control strategies to cope with typical problems encountered in practice.

2 Reduced-Order Dynamic Modelling

Wastewater processing by means of activated sludge is undoubtedly the most widespread sewage treatment practice. The activated sludge which forms the basis of the biological process is in fact a mixture of several microorganisms which under proper environmental conditions act as tiny bioreactors and transform the biodegradable pollutant, which constitutes their growth substrate, into protoplasm with energy being supplied by dissolved oxygen and carbon dioxide being released. The difficulty of modelling the whole chain of bioreactions lies in the complexity and variability of the microbial colony, with each species exhibiting a peculiar behaviour and interacting with all the others. Many mathematical models already exist, which take into account the full structure of the microbial dynamics and give a detailed picture of the bioreactions developing in the oxidation stage. In this area, credit ought to be given to Andrews [1-3], Busby and Andrews [4], and Olsson [7] for their pioneering work.

Hystorically, the first aspect to receive modelling attention was the biodegradation of carbonaceous substrate, globally measured by the "biochemical oxygen demand" (BOD), representing the amount of oxygen that the microorganisms require to metabolize the substrate. Later, the oxidation of nitrogenous wastes was considered and the first model describing the dynamics of nitrification was proposed by Poduska and Andrews [17]. Then the implications of the dissolved oxygen (DO) dynamics began to emerge. The coupling nature of DO and its potential role for control were highlighted by Olsson and Andrews [18] and Stenstrom and Andrews [19]. Presently, the detailed modelling of the interactions between the microbial colony and the complex substrate is well developed and very detailed, speculative models are available [20-25]. Given these premises and a much wider literature than that just quoted, one might wonder if there is really a need for yet another model. The answer lies in the complexity of these literature models, developed for speculative purpose, which make them unsuitable for the mathematical manipulations required in most control applications.

To fill the gap between modelling accuracy and control needs, a simplified model suitable for control applications was derived by the author [14,15] and is now briefly reviewed. It describes the dynamics of the main four process phases:

a) Biodegradation of carbonaceous BOD;
b) nitrification;

c) dissolved oxygen utilization;
d) sludge sedimentation.

Before introducing the complete continuous-flow process model, the dynamics of each subprocess is first derived for a batch situation.

2.1 Simplified Microbial Kinetics

This section summarizes the basic ideas underlying the development of a simplified kinetics involving substrate and activated sludge. The core of the model is represented by the interaction between carbonaceous BOD and the sludge biomass, which is globally modelled as "Mixed Liquor Suspended Solids" (MLSS). The kinetics that will be used in the sequel may be regarded as a simplified version of the well-known Monod dynamics of the substrate/biomass {S, X} pair, which is repeated here for convenience:

$$\text{Substrate:} \quad dS/dt = -\frac{1}{Y}\mu(S)\,X\;; \tag{1}$$

$$\text{biomass:} \quad dX/dt = \mu(S)\,X - K_d X \tag{2}$$

where the specific growth rate is defined as

$$\mu(S) = \frac{\hat{\mu} S}{K_s + S} \tag{3}$$

with $\hat{\mu}$ representing the maximum growth rate and K_s the half-maximum velocity constant. In contrast with the above Monod Eqs. (1)–(3), in the model proposed here the substrate/biomass interaction is viewed as an ecological prey/predator pair and modelled according to a modified Volterra-Leslie logistic equation (Maynard-Smith [26]). The resulting model, which will be referred to as SML, is now introduced. The two differential equations describing the dynamics of substrate and biomass are the following:

$$\text{Substrate:} \quad dS/dt = -K_b SX\;; \tag{4}$$

$$\text{biomass:} \quad dX/dt = K_c SX - K_m X^2/S \tag{5}$$

Where the positive constants $\{K_b, K_c, K_m\}$ are kinetic parameters thus defined:

K_b = substrate transformation rate coefficient (mg l^{-1} h^{-1});
K_c = biomass synthesis rate coefficient (mg l^{-1} h^{-1});
K_m = biomass endogenous metabolism rate coefficient (h^{-1}).

The basic assumptions under which the model was derived are the following:

1) The substrate uptake rate is proportional to the product of substrate and biomass (second-order kinetics).

2) The biomass growth rate is proportional to the substrate uptake rate, hence it has the same form as the substrate decay. In addition, a negative restraint term has been added to account for diminished growth due to substrate limitation. This term, expressed by the quadratic term X^2/S, represents the effect of endogenous metabolism and depends on substrate concentration.

Some qualitative features of the SML model (4)–(5) can now be easily established. Consider the critical biomass X^* given by the parabola

$$X^* = S^2K_c/K_m \tag{6}$$

It can be seen that Eq. (6) divides the first quadrant of the {S, X} plane in two regions, each with a clear biological meaning. Figure 1a shows that Eq. (6) partitions the {S, X} plane into a *synthesis* region (zone B) and an *endogenous metabolism* region (zone A). For any given substrate S, if $X < X^*$ the substrate/biomass system lies in the B zone, where synthesis is the prevailing metabolism and $dX/dt > 0$. Since dS/dt is always negative, any bioreaction originating in this region will tend towards the boundary (6), which is intersected with zero derivative ($dX/dS = 0$). Beyond this boundary the region with prevailing endogenous metabolism ($X > X^*$) is entered: both variables have negative derivatives and any bioreaction tends to the origin, which

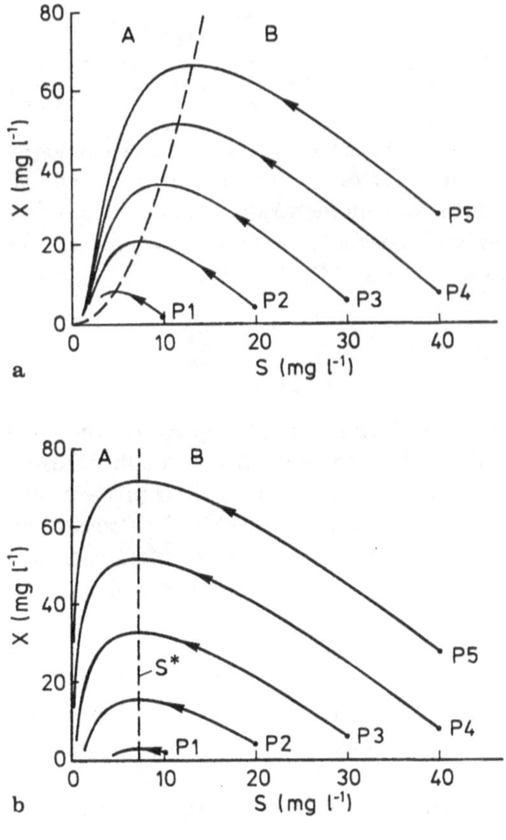

Fig. 1a and b. {S, X}-plane evolution of batch models. The boundary between the synthesis (Zone B) and endogenous methabolism (Zone A) regions is represented by the dashed curve: **a** SML model Eqs. (4), (5): the boundary is given by Eq. (6). **b** Monod model Eqs. (1)–(3): the boundary is given by Eq. (7)

is reached in an infinite time. Hence the model does not exhibit oscillations, which is an undesirable feature of the classical Leslie equation. In fact, any evolution originating in the first quadrant, with the exception of the X axis ($S = 0$), is bound to remain in this quadrant and will tend to the origin without ever reaching it. On the other hand, it cannot interesect either the S axis, since all the bioreactions tend to leave it, or the X axis as then S would tend to zero increasing the second term in (5) and dX/dt would tend to $-\infty$. Therefore, the X axis is an asymptote for all the possible batch evolutions.

Figure 1 b shows the equivalent trajectories produced by the Monod model (1)–(3), which exhibits a similar state space partitioning between synthesis and endogeneous metabolism. In this case the two regions are divided by a vertical line which depends only on the model parameters

$$S^* = \frac{K_s K_d}{\hat{\mu} - K_d} \tag{7}$$

There are several aspects which seem to favour model (4), (5) over the Monod kinetics. From the structural point of view, the endogenous metabolism is included in the basic kinetics and it is not required, as with Monod, to introduce an additional decay term for which the linear choice ($-K_dX$) is questionable. Furthermore, the SML model is linear in the parameters, which simplifies the numerical calibration as will be described in Sect. 3.

2.2 Dissolved Oxygen Kinetics

Aerobic bioprocessing of organic matter requires oxygen as a primary source of energy. It has been shown that this can become a limiting factor [27], but if it is plentiful enough to sustain growth then oxygen utilization is stoichiometrically related to biomass synthesis and endogenous metabolism. A key process variable, the "oxygen uptake rate" (OUR), can be defined on the basis of rate Eqs. (4), (5) as the sum of synthetic and maintenance activities

$$OUR = \gamma_s K_b SX + \gamma_e K_m X^2/S \tag{8}$$

where γ_s, γ_e are oxygen utilization coefficients. Equation (8) represents the total depletion rate of oxygen in the batch, which is in turn resupplied through diffusion from the gaseous phase. In fact, oxygen diffusion rate into the liquid is proportional to the difference between the dissolved oxygen concentration C and its saturation value C_{sat}, i.e. $K_La(C_{sat} - C)$ where K_La represents the oxygen mass transfer rate coefficient into the liquid phase. The complete dissolved oxygen balance is then

$$dC/dt = K_La(C_{sat} - C) - OUR \tag{9}$$

To gain further insight into the DO dynamics (9) and to simplify the subsequent parameter identification, a further relationship among DO balance parameters is now derived. Equations (4), (5) are solved for the terms SX and X^2/S and substituted into Eqs. (8), (9). Integrating over the whole duration of the batch (theoretically from

0 to ∞) with initial conditions {So, Xo} and under the assumption that the dissolved oxygen concentration is kept constant ($dC/dt = 0$), yields C_∞, the total amount of oxygen needed to metabolize the initial substrate So and for self-oxidation of the initial biomass Xo

$$C_\infty = \int_0^\infty K_La(C_{sat} - C)\,dt = -(\gamma_s + \gamma_e K_c/K_b) \int_{So}^{0} dS - \gamma_e \int_{Xo}^{0} dX \qquad (10)$$

In practice the substrate oxidation process takes a finite time, hence the left-hand-side integral is bounded. Reworking the integration limits of the right-hand-side integrals and solving yields

$$C_\infty = (\gamma_s + \gamma_e K_c/K_b)\,So + \gamma_e Xo \qquad (11)$$

If So and Xo are expressed in mg l^{-1} of oxygen and if the term $\gamma_e Xo$ includes all the self-oxidizing processes, then the total utilized oxygen C_∞ must equal the sum of the initial substrate So and the self-consumption term $\gamma_e Xo$. Therefore, the following must hold

$$C_\infty = So + \gamma_e Xo \qquad (12)$$

Equating Eqs. (11) and (12), the following relation among parameters is obtained

$$\gamma_s + \gamma_e K_c/K_b = 1\,. \qquad (13)$$

The interdependence of γ_s and γ_e is important in view of parameter calibration. In fact, as a consequence of Eq. (13) either γ_s or γ_e only need to be estimated.

2.3 Simplified Nitrification Kinetics

It is well known that nitrification occurs in two successive oxidation stages, with nitrite as an intermediate product. The complete chain of reactions was modelled by Poduska and Andrews [17], though more recently single-step nitrification models were proposed by Gujer and Erni [28] based on the notion that once the intermediate nitrite stage has been reached, the reaction proceeds until complete oxidation is achieved. Also, from the system-theoretical point of view, the $NH_4 \rightarrow NO_2$ stage is the slowest and therefore controls the overall reaction rate. Hence, the following single-stage model is proposed [15].

Ammonium-N: $$dS_{am}/dt = -\frac{\mu_n}{Y_a} X_n\,; \qquad (14)$$

Nitrate-N: $$dN_a/dt = -dS_{am}/dt\,; \qquad (15)$$

Nitrifiers: $$dX_n/dt = \mu_n X_n - K_n X_n \qquad (16)$$

where the rate expression μ_n takes the following composite form:

$$\mu_n = \hat{\mu}_n \frac{S_{am}C}{(K_{am} + S_a)(K_o + C)} \tag{17}$$

Stenstrom and Poduska [27] have demonstrated that the nitrification rate depends on both the amount of substrate and of dissolved oxygen available. It should also be noticed that the yield factor Y_a has not the same meaning of a conversion factor as in the case of carbonaceous BOD, since the nitrifiers take up a very limited amount of nitrogen from the ammonia pool, and act instead mainly as catalysts.

As far as oxygen utilization is concerned, the nitrification can be included in the DO balance already determined in Sect. 2.2 by simply adding a third term to the OUR expression (8) of the form

$$\gamma_n \mu_n X_n / Y_a \tag{18}$$

This factor is to represent the oxygen consumption for ammonium-N oxidation. Since the nitrifying bacteria mainly act as catalysts and do not use oxygen directly, a maintenance term is not included. Hence the full OUR expression for both carbonaceous and nitrogenous oxidation is the following

$$OUR_{cn} = \gamma_s K_b S X + \gamma_e K_m X^2/S + \gamma_n \mu_n X_n / Y_a \tag{19}$$

2.4 Sedimentation Dynamics

This section of the model is not directly concerned with biochemical reactions, but rather considers the double function of the secondary settler: separating the bioflocs from the liquid in order to produce a solid-free effluent, and thickening the biomass at the bottom of the settler to be recycled back into the aerator. Of these two actions, thickening is surely the most important and complex, whereas clarification may be regarded as a consequence of the former. In fact, although a clarification failure has an immediate impact on the process management, it is always the result of a thickening failure which would take longer to detect but whose implications are surely far more catastrophic.

Compared with the biochemical aspects of the process, sedimentation has received relatively less attention, and though the basic theory of flocculant suspensions is well established [29-31], the resulting partial differential equation models have often been neglected in favour of heuristic or empirical algebraic rules [32]. Sedimentation is a mass transfer process which can be modelled according to the solid flux theory [29, 31] to describe the subsidence of suspended solids through layers of differing concentrations. Consider the settler schematic diagram of Fig. 2 and the following diffusion equation

$$\frac{\partial X}{\partial t} = \frac{\partial F}{\partial z} = \frac{\partial F}{\partial X} \times \frac{\partial X}{\partial z} \tag{20}$$

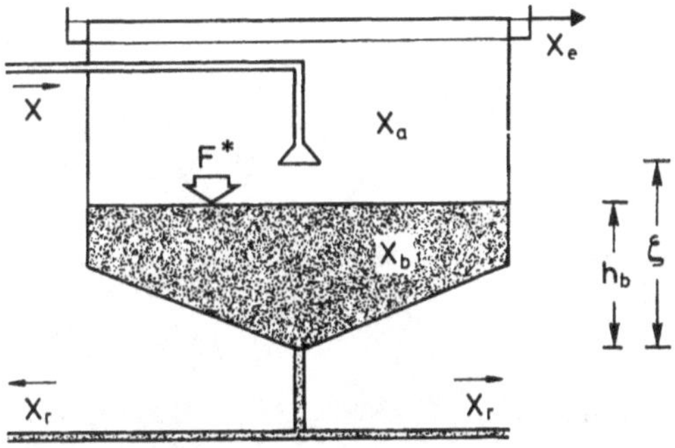

Fig. 2. Diagram of the secondary settler, showing the feed point at height ζ and the sludge discontinuity at height h_b between the clarification zone with concentration X_a and the compaction zone with concentration X_b

where F is the biomass solids flux of concentration X at height z from the thickener bottom. If Eq. (20) is approximated by finite differences, the following holds

$$\frac{\partial X}{\partial t} \simeq \frac{F_i - F_o}{\Delta z} \tag{21}$$

Equation (21) states that the rate of change of biomass concentration X through a horizontal layer of thickness Δz is equal to the difference between the incoming (F_i) and outgoing (F_o) fluxes divided by Δz. The motion of the bioflocs generating the solid flux is governed by two main forces: settling subsidence due to gravity and "bulk" flow due to sludge withdrawal from the bottom. Therefore the total flux can be expressed as

$$F = F(X, u) = Xv + Xu \tag{22}$$

where v and u are respectively the settling and bulk velocity of the bioflocs. Several analytical expressions have been proposed for the settling velocity, and the exponential and power-law approximations are the most widely used

exponential [31]: $v = n \exp(-aX)$; (23.1)

power law [32]: $v = nX^a$; (23.2)

inverse power law [33]: $v = \dfrac{a}{b + X^n}$ (23.3)

where the settling parameters {n, a} have different values and meaning in each of the three Eqs. (23). Severin et al. [90] have demonstrated with a series of experiments that the values of these parameters are very stable, even over a period of many seasons of operations. A family of solid flux curves obtained with Eq. (23.1) is shown in Fig. 3 as a function of solids concentration X and with bulk velocity u as a parameter. This can be expressed in terms of recycle and waste ratios r and w, assuming total solids capture at the thickener bottom

$$u = Q\frac{r + w}{A} \tag{24}$$

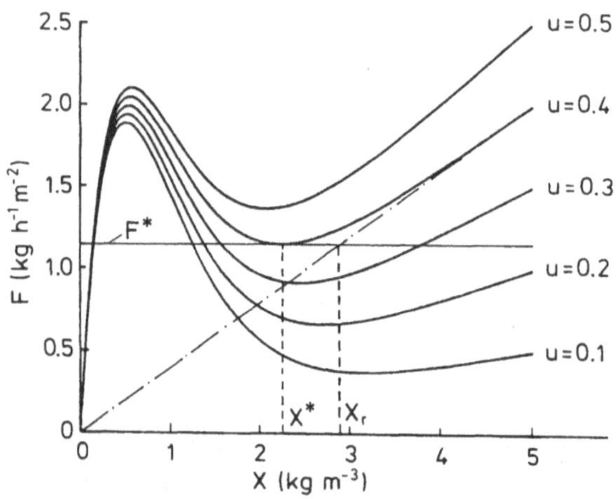

Fig. 3. Solid flux curves with bulk velocity u as a parameter, showing how the limiting quantities F* and X* and underflow concentration X_r are graphically related

where A is the thickener cross-section and Q the process flow rate. Figure 3 also shows that given a bulk velocity u, the underflow concentration X_r is uniquely determined by the solid flux established in the thickener. In fact the minimum flux F*(u) represents the maximum rate of solids transmission through the liquid/solid interface for a given clarifier geometry and sludge settling behaviour [29–33]. Excess solids which cannot be transferred to the bottom accumulate above the interface and eventually migrate upwards, leading to a clarification failure. In this sense clarification may be regarded as a consequential aspect of thickening. Conversely, the underflow solids concentration is not a dynamic variable but depends on the limiting flux and available mass above the interface. If the thickener is critically loaded, the total limiting flux AF*(u) passing through the thickening zone equals the underflow withdrawal. Equating the two fluxes, the underflow concentration is obtained

$$X_r = \frac{AF^*(u)}{Q(r + w)} \tag{25}$$

Figure 3 shows how X_r and F*(u) can be determined graphically, whereas for numerical computations the power law approximation (23.2) can be used. Expressing the total flux as

$$F(X, u) = nX^a + Xu \tag{26}$$

with $n > 0$ and $a < 0$, and letting $\partial F/\partial X = 0$, the limiting quantities X* and F* are determined

$$X^*(u) = \exp\left\{\frac{1}{a-1}\ln\left(-\frac{u}{na}\right)\right\} \tag{27}$$

$$F^*(u) = n\exp\left\{\frac{a}{a-1}\ln\left(-\frac{u}{na}\right)\right\} + u\exp\left\{\frac{1}{a-1}\ln\left(-\frac{u}{na}\right)\right\} \tag{28}$$

Now a solid mass balance around the thickener can be set up considering that the solids loading rate depends on the incoming concentration X and that the solids depletion rate from the underflow is governed by the limiting flux only

$$dM/dt = Q(1 + r)\,X - AF^*(u) \tag{29}$$

where M is the total mass of solids in the thickener and the underflow concentration is determined through Eq. (25), as shown in Fig. 3.

A slightly different approach has been proposed by Stehfest [33] resulting in a simple ordinary differential equation to describe the settler dynamics, and particularly that of the interface separating the clarified water from the thickened sludge (sludge blanket). Assuming again the settler scheme of Fig. 2, where the feed point and the position of the sludge blanket are shown, Stehfest reworked the original partial differential Eq. (20), taking into account the concentration above (X_a) and below (X_b) the sludge blanket, which is in fact a concentration discontinuity. Integrating Eq. (20) in the spatial variable z yields

$$\frac{d}{dt}\int_{z_a}^{z_b} X\,dz = F(X_b) - F(X_a) \tag{30}$$

If a discontinuity exists at an intermediate height $z_0 \in (z_a, z_b)$ then Eq. (30) yields

$$dz_0/dt = -\frac{F(X_a) - F(X_b)}{X_a - X_b} \tag{31}$$

From Eqs. (30) and (31) a mass balance above and below the sludge blanket can be written under the assumption that the two zones above and below the discontinuity have homogeneous concentrations X_a and X_b. After a number of intermediate manipulations, mainly due to continuity constraints and the propagation of rarefaction waves from the clarifier bottom, the following lumped-parameter model is obtained:

Solids concentration above the interface (X_a):

$$dX_a/dt = \frac{1}{\zeta - h_b}\,[F_i(X) - F(X_a)]\,; \tag{32}$$

solids concentration below the interface (X_b):

$$dX_b/dt = \frac{1}{h_b}\left\{F(X_a) - F(X_o) - \frac{X_b - X_a}{X_- - X_a}\,[F(X_a) - F(X_-)]\right\}; \tag{33}$$

sludge blanket height (h_b):

$$dh_b/dt = \frac{F(X_a) - F(X_-)}{X_- - X_a} \tag{34}$$

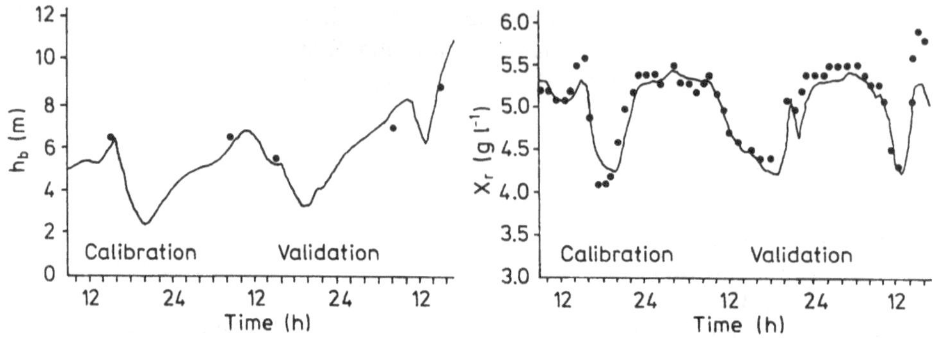

Fig. 4. Experimental performance of Stehfest model [33] Eqs. (32)–(34). The sludge blanket height (h_b) and underflow sludge concentration (X_r) are shown. The data in the first 24-h period are used for calibration and those in the following 24-h for validation. Reproduced by permission of the Institute of Measurement and Control on behalf of Stehfest (33)

where
$X_- = \min(X_b, X_a^t)$
$X_o = \max(X_b, X_*)$
ζ = feed height above clarifier bottom (m).

Model (32)–(34) was practically calibrated with measurements from municipal wastewater treatment plants and the agreement between experimental data and model predictions are shown in Fig. 4.

So far the thickening function of the settler has been considered. As far as clarification is concerned, empirical relations have been used almost invariably [31, 32] but recently a dynamic double black-box approach was proposed by Olsson and Chapman [34] to model the effluent suspended solids. It is based on the notion that the clarification response to influent step inputs varies depending on the sign of the step. In fact, changes in effluent suspended solids following a step increase were rapid and possibly with some overshoot, whereas a step decrease produced an exponential decline in turbidity. Two separate linear models were adopted, one for each direction of the flow rate change:

Positive flow-rate change:

$$d^2X_e/dt^2 + a_1\, dX_e/dt + a_2X_e = b_1\, dQ/dt + b_2Q\ ; \tag{35.1}$$

negative flow-rate change:

$$dX_e/dt + a_1X_e = b_1Q\ , \tag{35.2}$$

where X_e is the concentration suspended solids in the clarifier effluent and $\{a_1, a_2, b_1, b_2\}$ are model parameters that can be adjusted on-line during process operation. Combining the two Eqs. (35.1) and (35.2) a good agreement between observed and predicted effluent suspended solids has been achieved by the authors, as shown in Fig. 5 from [34].

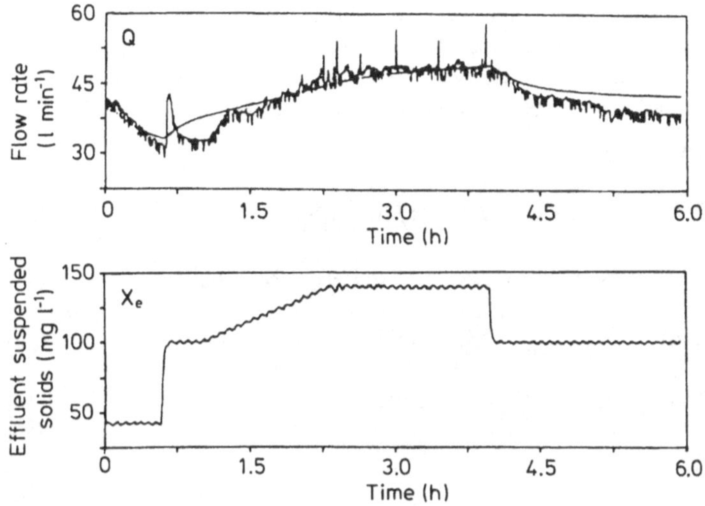

Fig. 5. Double linear approximate clarifier model Eqs. (35): flow rate disturbance performance [34]. Published by the International Association on Water pollution Control (IAWPRC) in conjunction with Pergamon Press

2.5 Continuous-Flow Model

Having considered the batch kinetics of the main process streams, the complete continuous-flow model can be easily obtained adding the appropriate input-output transport terms. Hereafter, the bioreactor is considered to be perfectly mixed so that the concentration of each component is spatially homogeneous. The opposite to the completely mixed arrangement is the plug flow reactor [82], where no transversal mixing takes place. Practical plants lie in between these two extremes, being closer to either kind depending on tank geometry and flow patterns.

Figure 6 depicts the process scheme of a completely-mixed activated sludge treatment composed of an aerator of volume V and a secondary settler with cross section A. The recycle and waste streams expressed as fractions of the process flow rate Q are also shown. S_i is the incoming organic load usually expressed as BOD, whereas the biomass in the aerator is referred to as mixed liquor suspended solids (MLSS).

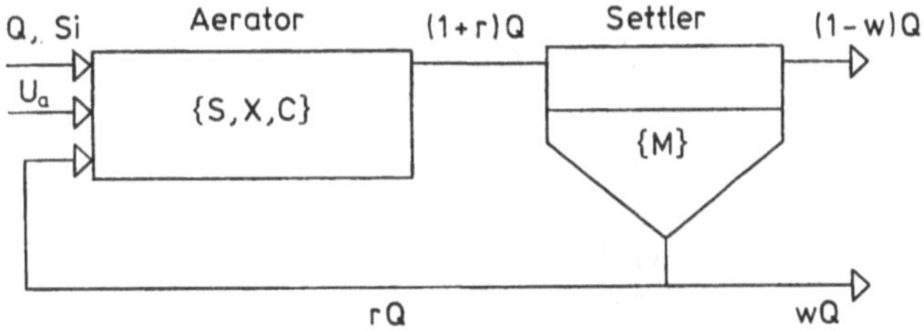

Fig. 6. Schematic diagram of a continuous-flow completely mixed activated sludge process

A straightforward hydraulic mass balance yields the necessary transport terms to supplement the kinetics already established by Eqs. (4), (5), (9) whereas the settler mass balance of Sect. 2.4 was already derived for a continuous-flow process. Then, the continuous-flow model comprises the following equations:

a) *Differential equations:*

BOD: $$dS/dt = -K_b SX - q(1 + r)S + qS_i ; \tag{36}$$

MLSS: $$dX/dt = K_c SX - K_m X^2/S - q(1 + r)X + q_r X_r ; \tag{37}$$

DO: $$dC/dt = K_L a(C_{sat} - C) - OUR - q(1 + r)C + qC_i ; \tag{38}$$

settler MASS: $$dM/dt = Q(1 + r)X - AF^*(u) . \tag{39}$$

b) *Auxiliary equations:*

Underflow MLSS: $$X_r = \frac{AF^*(u)}{Q(r + w)} ; \tag{40}$$

oxygen uptake rate: $$OUR = \gamma_s K_b SX + \gamma_e K_m X^2/S \tag{41}$$

where $q = Q/V$ is the dilution rate and C_i the incoming dissolved oxygen. It should be stressed that all the above Eqs. (36)–(41) deal with concentrations, with the exception of Eq. (39), which represents a global balance and hence involves global quantities such as clarifier cross section A and total incoming flow Q. Equations (36)–(41) constitute the basic dynamic model for carbonaceous BOD oxidation from which estimation and control algorithms are now derived. The typical numerical values of the parameters appearing in these model equations are shown in Table 1. They were

Table 1. SML model [14, 15] parameter ranges

Parameter	Value	Units	Ref.
K_b	$5.2 \times 10^{-4} \div 7.5 \times 10^{-4}$	$mg^{-1}\,l\,h^{-1}$	14, 52)
K_c	$4.7 \div 10^{-4} \div 1.5 \times 10^{-3}$	$mg^{-1}\,l\,h^{-1}$	14, 52)
K_m	$1.2 \div 10^{-5} \div 5.2 \times 10^{-4}$	h^{-1}	14, 52)
γ_s	$0.51 \div 0.71$	—	15, 52)
γ_e	$0.24 \div 0.36$	—	15, 52)
$K_L a$	$0.14 \div 0.27$	h^{-1}	15)
Y_a	$0.05 \div 0.067$	—	17)
$\hat{\mu}_n$	$0.04 \div 0.058$	h^{-1}	17)
K_n	$0.0025 \div 0.004$	h^{-1}	17)
K_{am}	$9.5 \div 11.5$	$mg\,l^{-1}\,h^{-1}$	17)
K_o	$0.5 \div 1.5$	$mg\,l^{-1}\,h^{-1}$	27)
γ_n	$0.33 \div 0.46$	—	15)
n	$3.76 \div 5.2$	—	31–33)
a	$-2.0 \div -2.25$	—	31–33)

obtained either from literature value or from specific experiments, as described in the following section.

3 Model Identification

Identifying a dynamic model entails the determination of the numerical values of its parameters so that the model response reproduces as closely as possible the available experimental data. The most widely used "discrepancy measure" between model output and experimental data is the sum of squared differences

$$E_{sx} \equiv \sum_{i=1}^{N} [S(i) - S_{exp}(i)]^2 + [X(i) - X_{exp}(i)]^2 \tag{42}$$

where $\{S_{exp}(i), X_{exp}(i); i = 1, 2, \dots, N\}$ are experimental data of substrate and biomass. It must be stressed that E_{sx} is a function of model parameters and as such its actual form depends on the model equations being used.

The determination of the parameter values of microbial kinetics has represented a challenging endeavour ever since such mathematical models were first introduced. Nonlinear equations, such as Monod kinetics Eqs. (1)–(3), together with data scarcity and inaccuracy, generated new identification problems and ad hoc estimation procedures were devised to deal with these difficulties, as described by early contributions such as Heineken et al. [35] and Naito et al. [36]. This section reviews most of the research developed in the last decade for the identification of activated sludge kinetic parameters. However, before undertaking the actual parameter identification it is necessary to assess whether the model equations are such that their parameters can be identified from experimental data. This intrinsic model property is termed identifiability and its assessment is a necessary prerequisite to any practical parameter identification endeavour.

3.1 SML Model Identifiability

The difficulties encountered in the identification of the Monod dynamics have been extensively reported in the literature [35–42]. They arise mainly from the close correlation between $\hat{\mu}$ and K_s, as pointed out particularly by Holmberg and Ranta [42]. In fact it can be shown that the estimation error E_{sx} defined by Eq. (42), when combined with the Monod kinetics (1)–(3), produces a narrow and elongated "valley" around the minimum corresponding to the best fitting model parameters. An awkward shape of the estimation error E_{sx} is likely to cause severe numerical problems to the identification algorithm.

Since the SML model (4), (5), (9) was derived independently of the Monod kinetics and has a quite different structure, it should not be expected to encounter the same kind of difficulties and the results obtained for the Monod model cannot be transferred directly into this context. Therefore a new identifiability analysis is now carried out in two steps: first, a general test is performed to assess the model identifiability from a theoretical point of view, then the more practical aspects are considered.

For the first part, the basic notation and definitions introduced by Di Stefano and Cobelli [43)] are used throughout. Consider a dynamic system described by the vector differential equations

$$d\mathbf{Z}(t, \mathbf{P})/dt = \mathbf{f}(\mathbf{Z}(t, \mathbf{P}), t, \mathbf{P}) \tag{43}$$

$$\mathbf{Y}(t, \mathbf{P}) = \mathbf{g}(\mathbf{Z}(t, \mathbf{P}), t, T) \tag{44}$$

with initial conditions

$$\mathbf{Z}_0 = \mathbf{Z}(t_0, \mathbf{P}) \tag{45}$$

where $\mathbf{Z} \in R^n$, $\mathbf{Y} \in R^q$, $\mathbf{P} \in R^p$ are the state, output, and parameter vectors respectively, and $\mathbf{f}(.)$ and $\mathbf{g}(.)$ are vector-valued nonlinear functions of consistent dimensions. The system is said to be parameter identifiable on the interval (0, T) if there exists a unique solution to the parameter vector $\mathbf{P}$ from Eqs. (43)–(45). The theoretical identifiability, sometimes referred to as structural identifiability, can be assessed through the Pohjampalo test [44)], which states that the system (43)–(45) is identifiable over the interval (0, T) if there exists a unique solution to the set of equations

$$\mathbf{g}^{(k)}(\mathbf{Z}(t_0, \mathbf{P})) = \mathbf{Z}_k(0) \qquad k = 0, 1, 2, \ldots \tag{46}$$

where $\mathbf{g}^{(k)}$ is the k-th derivative of the vector function $\mathbf{g}(.)$ and $\mathbf{Z}_k(0)$ is the k-th derivative of the state vector $\mathbf{Z}$ evaluated at $t = 0$. This test assumes perfect continuous measurements along the interval (0, T) and in principle relates the identifiability to the reconstructibility of the initial state $\mathbf{Z}_0$.

Therefore it does not take into account the state evolution in the interval (0, T) and consequently it disregards any situation in which the state variables and/or the available data may be such as to render the identification less reliable. Thus, the Pohjampalo test should be regarded just as a preliminary stage of identifiability assessment, before applying more realistic tests which take into account the actual experimental setting. Assuming that all the state vector $\mathbf{Z} = [S, X, C]^T$ is accessible (i.e. $\mathbf{Y} = \mathbf{Z}$ and $\mathbf{f} = \mathbf{g}$) and indicating with $\{S_0, X_0, C_0\}$ the state variables at $t = 0$ and with $\{S_1, X_1, C_1, X_2\}$ their first and second derivatives at $t = 0$, the following quantities are obtained

$$S_1 = -K_b S_0 X_0 \tag{47.1}$$

$$X_1 = K_c S_0 X_0 - K_m X_0^2/S_0 \tag{47.2}$$

$$C_1 = K_L a(C_{sat} - C_0) - \gamma_s K_b S_0 X_0 - \gamma_e K_m X_0^2/S_0 \tag{47.3}$$

$$X_2 = K_c(S_1 X_0 + S_0 X_1) - K_m\left(\frac{2X_0X_1S_0 - X_0S_1}{S_0^2}\right) \tag{47.4}$$

According to the Pohjampalo test, the model (4), (5), (9) is identifiable if the parameters $\{K_b, K_c, K_m, \gamma_e\}$ can be obtained by solving Eqs. (47). The K_b parameter can be obtained independently solving Eq. (47.1) and then γ_e is obtained from Eq. (47.3), making use of Eq. (47.2)

$$\gamma_e = \frac{C_1 - K_L a(C_{sat} - C_0) + K_b S_0 X_0}{X_1} \tag{48}$$

The two remaining parameters K_c and K_m can be obtained from the linear system Eq. (47.2) and Eq. (47.4) provided that the matrix

$$D = \begin{vmatrix} X_0 S_0 & -\dfrac{X_0^2}{S_0} \\ X_0 S_1 + X_1 S_0 & -\dfrac{2X_1 X_0 S_0 - X_0^2 S_1}{S_0^2} \end{vmatrix} \tag{49}$$

is non-singular. It is easy to show that singularity occurs if and only if

$$2\frac{X_0}{S_0} = \frac{X_1}{S_1} \tag{50}$$

Condition (50) will never hold in a batch situation. In fact, whereas X_0 and S_0 are always positive, the derivatives at $t = 0$ have opposite signs since the substrate can only decrease ($S_1 < 0$) whereas the biomass can only increase ($X_1 > 0$). Since the two members of (50) can never have the same sign, this equality never holds and therefore matrix D cannot become singular, thus assuring a unique solution for K_c and K_m. The above reasoning proves that all the parameters $\{K_b, K_c, K_m, \gamma_e\}$ are identifiable whenever $S_0 > 0$ and $X_0 > 0$, which is the only realistic situation.

3.2 Parametric Sensitivity

As already stated, the Pohjampalo test is a minimal guarantee that the system is indeed identifiable. More practical considerations are now required to ensure practical identification with limited experimental data. Sensitivity analysis provides a quantitative assessment of the extent to which parameter errors influence model response. Consider again the nonlinear vector dynamic system Eq. (43)

$$d\mathbf{Z}/dt = \mathbf{f}(\mathbf{Z}, \mathbf{P}, t) \tag{51}$$

where $\mathbf{Z} \in R^n$ is the state vector and $\mathbf{P} = \{p_i;\ i = 1, \dots, p\} \in R^p$ is the parameter vector. The sensitivity S_z^i with respect to the i-th parameter p_i is then defined as the incremental variation of the state vector Z caused by an incremental variation of the i-th parameter p_i, namely

$$S_z^i = \frac{\partial Z}{\partial p_i} \tag{52}$$

A sensitivity system can be associated to the given system (51) simply by applying the definition (52) to yield the following sensitivity-generating system with respect to p_i

$$dS_z^i/dt = \left(\frac{\partial f}{\partial Z}\right)_n S^i + \left(\frac{\partial f}{\partial p_i}\right)_n \tag{53}$$

where the terms in parentheses are computed along the trajectory with nominal parameter values $p_i^{(n)}$. These results, mainly due to Perkins [45], are useful for practical identifiability assessment. In fact the evolution of the sensitivity system (53) provides information on the practical identifiability of the system from a given set of measurement. In fact, as pointed out by Holmberg [40] and later by Vialas et al. [46], the sensitivities associated to the Monod kinetics (1)–(3) show that this model takes full advantage of the information contained in the data only during a short part of a typical

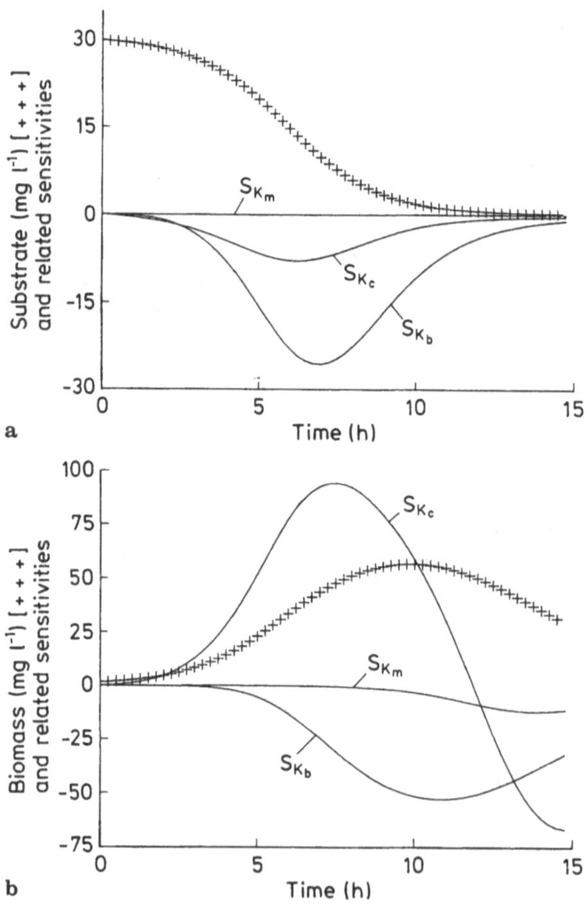

Fig. 7a and b. Trajectory sensitivity of the SML batch model: **a** Substrate and related sensitivities, **b** biomass and related sensitivities

batch experiment and that large estimation errors are generated. The trajectory sensitivity analysis is now applied to the reduced-order model (4), (5) in order to assess its practical identifiability. The system matrices appearing in the sensitivity system (53) are the following

$$\partial f/\partial Z = \begin{vmatrix} -K_bX & -K_bS \\ K_cX + K_mX^2/S^2 & K_cS - 2K_mX/S \end{vmatrix} \tag{54}$$

$$\partial f/\partial p = \begin{vmatrix} -SX & 0 & 0 \\ 0 & SX & -X^2/S \end{vmatrix} \tag{55}$$

The trajectory sensitivities obtained by substitution of matrices (54), (55) into Eq. (53) are shown in Figs. 7a and 7b and can be compared with the corresponding Monod curves of Figs. 8a and 8b. By inspection two main differences can be noticed:

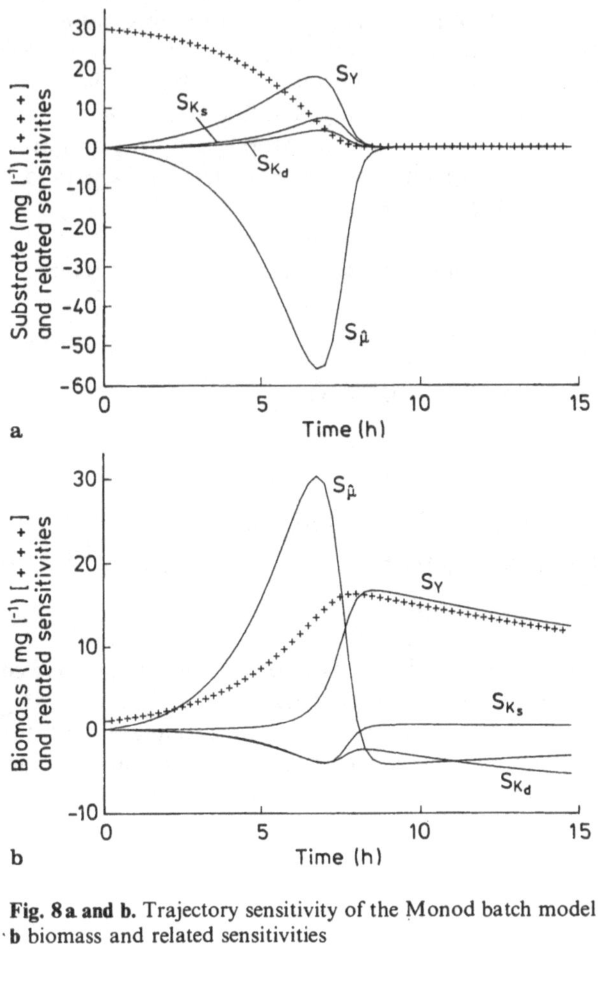

Fig. 8a and b. Trajectory sensitivity of the Monod batch model: **a** Substrate and related sensitivities, **b** biomass and related sensitivities

1) In the SML model the sensitivities of S are not all proportional, as in the Monod case, thus implying limited parametric interdependence. As noted by Holmberg [40], it is the intercorrelation of $\hat{\mu}$ and K_s that makes their identification difficult. With the SML model, this problem does not arise.
2) The biomass sensitivities of the SML model do not exhibit a sharp peak as with the Monod kinetics [40, 46]. Therefore this model is expected to take better advantage of the information provided by the entire data set.

It can be concluded that numerical calibration of model (4), (5) from batch data is feasible and numerically reliable. This conclusion is independent of the particular numerical method being used since it refers solely to the structural properties of the model.

3.3 Practical Model Identification

The preceding Sects. 3.1 and 3.2 have assessed the SML model identifiability, in other words it was demonstrated that the structure of the batch model Eqs. (4), (5) is such that their parameters can be determined on the basis of a suitable number of experimental data. Now, the numerical accuracy of the estimates will be assessed using different numerical methods. In particular, it is important to check how the model parameters interrelations affect the estimation reliability. The first attempt to estimate the parameters of the SML continuous-flow model was proposed by the author [14] using the error equation approach and taking advantage of the fact that this model, unlike the Monod kinetics, is linear in the parameters.

Let $\{S_{exp}(i), X_{exp}(i);\ i = 1, 2, \ldots, N\}$ be a collection of substrate and biomass experimental data. If they are substituted in the process Eqs. (36), (37) the following set of equations are obtained

BOD:

$$dS_{exp}(i)/dt = -K_b S_{exp}(i)\, X_{exp}(i) - q(1 + r)\, S_{exp}(i) + qS_i + \varepsilon_1(i) \quad (56)$$

MLSS:

$$\begin{aligned} dX_{exp}(i)/dt &= K_c S_{exp}(i)\, X_{exp}(i) - K_m X^2_{exp}(i)/S_{exp}(i) \\ &- q(1 + r)\, X_{exp}(i) + qrX_r + \varepsilon_2(i) \\ &\text{for}\quad i = 1, 2, \ldots, N \end{aligned} \quad (57)$$

where $\varepsilon_1(i)$ and $\varepsilon_2(i)$ are the equation errors induced by imperfect matching between the experimental data and the model equations. Supposing that the plant operating factors $\{q, r, X_r, S_i\}$ are known, the best fit of experimental data is obtained with the parameter set $\{\hat{K}_b, \hat{K}_c, \hat{K}_m\}$ which minimizes the sum of the squared equation errors $\varepsilon_1^2(i)$ and $\varepsilon_2^2(i)$. The parameter identification then gives rise to the following optimization problem:

$$\min_{K_b, K_c, K_m} \sum_{i=1}^{N} \varepsilon_1^2(i) + \varepsilon_2^2(i) \quad (58)$$

or, equivalently,

$$\min_{K_b, K_c, K_m} \sum_{i=1}^{N} \{[(dS_{exp}(i)/dt) - (dS/dt)]^2 + [(dX_{exp}(i)/dt) - (dX/dt)]^2\} \quad (59)$$

In principle this approach seeks to minimize the sum of quadratic errors between the experimental and model derivatives rather than the process variables S and X. Taking advantage of the parametric linearity of Eqs. (56), (57), the optimization problem (59) is in fact a linear least-square problem and therefore can be solved analytically through the intermediate quantities

$$A_1 = \sum_{i=1}^{N} S^2_{exp}(i)\, X^2_{exp}(i) \quad (60)$$

$$A_2 = \sum_{i=1}^{N} X^4_{exp}(i)/S^2_{exp}(i) \quad (61)$$

$$A_3 = \sum_{i=1}^{N} X^3_{exp}(i) \quad (62)$$

$$Y_1 = \sum_{i=1}^{N} [S_{exp}(i)\, X_{exp}(i)\, (dS_{exp}(i)/dt) + q(1 + r)\, X^2_{exp}(i)\, S_{exp}(i) - qX_rX_{exp}(i)\, S_{exp}(i)] \quad (63)$$

$$Y_2 = \sum_{i=1}^{N} \left\{ q(1 + r) \frac{X^3_{exp}(i)}{S_{exp}(i)} - [qrX_r + (dX_{exp}(i)/dt)] \frac{X^2_{exp}(i)}{S_{exp}(i)} \right\} \quad (64)$$

then the following estimates are obtained:

$$\hat{K}_b = \sum_{i=1}^{N} [qS_iS_{exp}(i)\, X_{exp}(i) - q(1 + r)\, S^2_{exp}(i)\, X_{exp}(i) - (dS_{exp}(i)/dt)\, S_{exp}(i)\, X_{exp}(i)]\, A_1^{-1} \quad (65)$$

$$\hat{K}_c = (Y_1A_2 - A_3Y_2) \times (A_1A_2 - A_3^2)^{-1} \quad (66)$$

$$\hat{K}_m = (Y_2A_1 - A_3Y_1) \times (A_1A_3 - A_3^2)^{-1} \quad (67)$$

This procedure, though conceptually very simple, conceals some pitfalls. First, the equation errors $\varepsilon_1(i)$ and $\varepsilon_2(i)$ are nonlinear functions of the data errors and therefore can produce biased estimates. Secondly, the derivatives of the experimental quantities $dS_{exp}(i)/dt$ and $X_{exp}(i)/dt$ appearing in the estimator Eqs. (60)–(67) should be approximated with a suitable numerical technique to reduce the error amplification typical of numerical derivatives. In this application, the raw data were smoothed using least-square approximating splines [47] and both the order of the spline and their knots locations were used as smoothing parameters to reduce the experimental errors. This technique, described in details in [14], gives adequate results as shown in Fig. 9 where a set of 17 hourly BOD data was used to calibrate the model. Several sets of experimen-

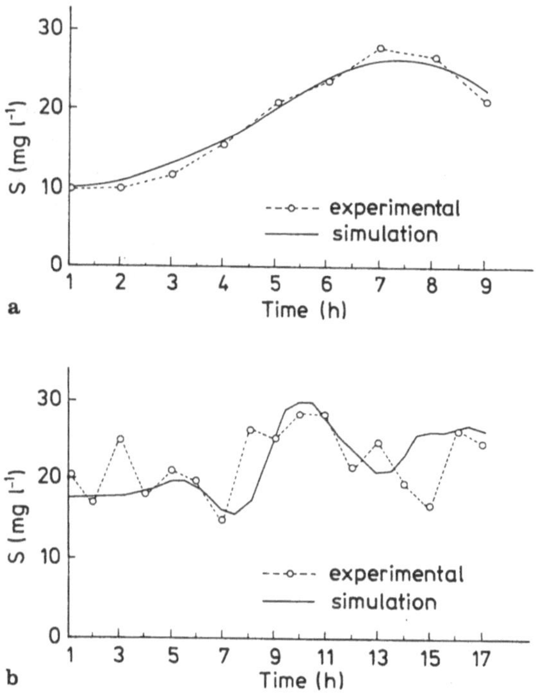

Fig. 9a and b. SML model response to a varying BOD input. Model calibrated using the estimator Eqs. (60)–(67) with splines approximation of the derivative [14]: **a** Municipal medium-scale plant, **b** pilot-plant

Table 2. SML model parameter obtained from operating plants [14]

Parameter	Value	Units
Municipal medium-scale plant:		
K_b	7.413×10^{-4}	$mg^{-1}\,l\,h^{-1}$
K_c	1.482×10^{-3}	$mg^{-1}\,l\,h^{-1}$
K_m	5.213×10^{-4}	h^{-1}
Pilot plant:		
K_b	6.134×10^{-4}	$mg^{-1}\,l\,h^{-1}$
K_c	1.935×10^{-3}	$mg^{-1}\,l\,h^{-1}$
K_m	5.023×10^{-4}	h^{-1}

tal data were taken from a pilot-plant operated by the municipality of Modena and from several medium-scale completely mixed plants operated by the municipality of Florence. The parameter calibration results are summarized in Table 2. Their values confirm that the efficiency of the plant is proportional to its size. In fact, recalling that K_b and K_c are related to the ability of the biomass to degrade organic substrate, the medium-scale plants exhibit a higher value of this parameter. On the other hand, efficiency computation carried out on the same plants showed that while the efficiency of the pilot plant was only 80%, that of the medium-size plants was above 90%. Further, the ratio K_b/K_c is both conceptually and numerically very close to the "Yield Factor" Y used in the Monod kinetics to indicate the transformation efficiency of substrate. From the values of Table 2, the ratio K_b/K_c is 0.5 for the medium-size plants and only 0.33 for the pilot plant, again confirming that size affects treatment efficiency.

The substrate/biomass continuous-flow model (36), (37) was obtained adding the pertinent hydraulic mass balance terms to the batch kinetics already specified as Eqs. (4), (5). Since these terms can be estimated independently, it may be practical to calibrate the kinetic part independently from the hydraulic balance, which can be determined by other simple means. In other words, the kinetic parameters $\{K_b, K_c, K_m\}$ can be determined by analyzing a suitable batch reaction carried out by the already acclimated biomass, irrespective of the hydraulic operating conditions. This approach represents a viable alternative to the previous one, as the minimization problem now can be restated as follows

$$\min_{\{K_b, K_c, K_m\}} E_{sx} = \sum_{i=1}^{N} [S(i) - S_{exp}(i)]^2 + [X(i) - X_{exp}(i)]^2 \tag{68}$$

with Eqs. (4), (5) as optimization constraints:

$$dS/dt = -K_b SX \tag{4}$$

$$dX/dt = K_c SX - K_m X^2/S \tag{5}$$

As already pointed out, a typical difficulty in calibrating microbial growth models is caused by the interrelation between kinetic parameters. This is particularly apparent with the pair $\{\hat{\mu}, K_s\}$ of the Monod kinetics, as thoroughly discussed by Holmberg and Ranta [42]. In fact, if these two parameters are varied so as to obtain a constant value E^*_{sx} of the functional (68), then very narrow and elongated closed contours are obtained in the $\{\hat{\mu}, K_s\}$ plane. As a result, the optimal point situated at the center of

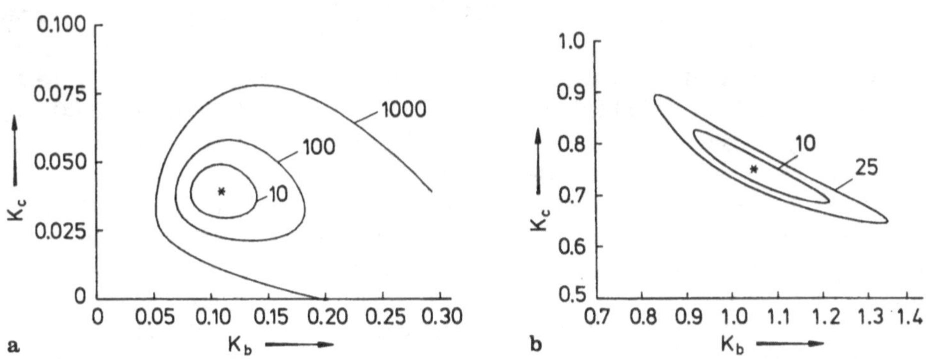

Fig. 10a and b. Error level contours of the SML model as a function of K_b and K_c: **a** E_{sx} from Eq. (68) with both variables S and X available, **b** E_c from Eq. (69) with only dissolved oxygen C available

this narrow "valley" is reached with great difficulty by numerical optimization algorithms, which may converge poorly, produce unreliable estimates or fail altogether. It is therefore appropriate to study the shape of the cost functional (68) when connected to model (4), (5) and ensure that such a situation is avoided. Figure 10a depicts the level contours of E_{sx} as a function of $\{K_b, K_c\}$ parameters, showing that the elongation, caused by the interrelation between parameters, though present, is quite acceptable. Further, to avoid the numerical derivative problem non-gradient methods were used such as the flexible polyhedron algorithm [48,49] in which the basic search procedure has been improved by optimizing the pattern search adjustments [50]. This enables the algorithm to cope successfully with the "narrow valley" problem resulting in a highly efficient search. In fact, the modified algorithm can adjust the search parameters to the shape of the functional, directing the search along the trough main direction and selecting a projection length which is optimal for each particular step. As a practical result, if a mixed liquor sample is taken out of the plant oxidation tank and its oxygen consumption progress is analyzed in a stand-alone batch, the substrate/biomass interaction will be the same as in the previous continuous-flow arrangement, since the microbial colony is already acclimated to that particular substrate, but with obvious practical advantages.

The identification problem becomes more difficult if not all the biological quantities are experimentally available. A practical case is when only dissolved oxygen data are available instead of substrate and biomass data. Then, the dissolved oxygen dynamics (9) has to be included in the model and consequently the optimization problem (68) becomes

$$\min_{\{K_b, K_c, K_m, \gamma_e\}} E_c = \sum_{i=1}^{N} [C(i) - C_{exp}(i)]^2 \tag{69}$$

with the following model as optimization constraint:

$$dS/dt = -K_b SX \tag{4}$$

$$dX/dt = K_c SX - K_m X^2/S \tag{5}$$

$$dC/dt = \gamma_s K_b SX + \gamma_e K_m X^2/S \tag{70}$$

Figure 10b shows that this new arrangement produces a narrower trough in the E_c contour portrait. Now, the identification problem (69) is that of fitting model (4), (5), (70) with only oxygen measurements available. Although identification is numerically more difficult, nonetheless the structural properties previously demonstrated still hold and the problem (69) has indeed a practical solution. This is precisely the situation encountered when determining the BOD content of a sample through the well-known respiration test, as described for example by Bhatla and Gaudy [51]. This kind of curve (usually termed "exertion curve") exhibits two peculiar features: the incubation delay, caused by the initially slow growth of bacteria, and the "plateau" when the microbial colony becomes mature and endogenous metabolism prevails over synthesis. The pertinent model for oxygen utilization is precisely Eq. (70) as shown in [52], because in this kind of experiment no additional oxygen is supplied throughout the test and

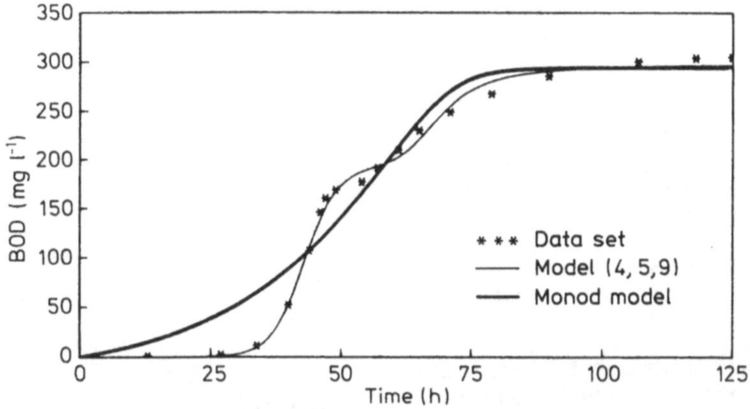

Fig. 11. Fitting batch oxygen respiration data [51] with the Monod model Eqs. (1), (2), (71) (thick line) and SML reduced-order model Eqs. (4), (5), (70) (thin line) [52]

therefore the diffusion term $K_La(C_{sat} - C)$ of Eq. (9) must be omitted. In other words, the measured quantity is indeed the integral of Eq. (70) over the test period (usually lasting five days). Both the SML model (4), (5), (70) and the classical Monod model were calibrated using the experimental BOD respiration data from [51]. For this purpose, the Monod model was rewritten as

$$dS/dt = -\frac{1}{Y}\mu(S)\,X \tag{1}$$

$$dX/dt = \mu(S)\,X - K_dX \tag{2}$$

$$dC/dt = \gamma_s \frac{1}{Y}\mu(S)\,X + \gamma_e K_d X \tag{71}$$

where $\mu(S)$ is again provided by Eq. (3). Figure 11, where the two model responses are compared, shows that the SML model (4), (5), (70) fits the experimental data better than the Monod model (1), (2), (71). The numerical values of the model parameters are grouped in Table 3. It can be noticed that the Monod model fails to reproduce the plateau adequately and that the initial incubation delay is grossly missed. As a general remark, it appears that the sigmoid response of the Monod kinetics has poor flexibility and it is not amenable to fitting the respiration curve properly. This is also confirmed by the massive estimation error in the respiration coefficients γ_s and γ_e, which are almost unidentifiable. Some other parameters too show a considerable estimation inaccuracy, notably K_s. This result agrees with Holmberg [40] who reported estimation experiments in which the standard deviation of the estimated K_s is of the same order of magnitude as the parameter itself. This test further demonstrates the difficulties in using the Monod kinetics and the advantage of the simpler SML model.

Table 3. BOD exertion curve model parameters [52]

Parameter	Range of values	Units
1) Monod model:		
$\hat{\mu}$	0.343 ± 0.082	h^{-1}
K_s	15.696 ± 5.232	$mg\,l^{-1}$
Y	0.583 ± 0.151	—
K_d	0.283 ± 0.707	h^{-1}
γ_s	0.547 ± 2.293	—
γ_e	0.777 ± 3.238	—
S_0	298.03 ± 43.5	$mg\,l^{-1}$
X_0	1.63 ± 0.89	$mg\,l^{-1}$
2) SML model:		
K_b	$(7.261 \pm 0.658) \times 10^{-4}$	$mg^{-1}\,l\,h^{-1}$
K_c	$(10.784 \pm 0.287) \times 10^{-4}$	$mg^{-1}\,l\,h^{-1}$
K_m	$(8.973 \pm 2.569) \times 10^{-4}$	h^{-1}
γ_s	0.511 ± 0.076	—
γ_e	0.241 ± 0.026	—
S_0	299 ± 14.3	$mg\,l^{-1}$
X_0	3.3 ± 0.54	$mg\,l^{-1}$

3.4 On-line Parameter Identification

So far off-line methods have been considered for the identification of the microbial kinetics. For many biotechnological control applications, though, it is important to obtain real-time parameter estimates as the process unravels. The most widely used approach is based on the "extended Kalman filter" (EKF) approach coupled with the Monod kinetics (1)–(3). The general structure of the EKF algorithm is described in a number of outstanding books, notably Jazwinski [53] and Gelb [54], and an early biotechnical application was reported by Svrcek et al. [55]. The application of the EKF algorithm to the microbial kinetics is particularly relevant because both the process variables (substrate and biomass) and the related parameters $\{\hat{\mu}, K_s, K_d, Y\}$ are usually unknown and/or time-varying. The EKF approach considers the unknown parameters as additional process variables and seeks to estimates the extended vector (hence the EKF name) of variables by filtering the differences between model outputs and experimental measurements as soon as these become available. The application of EKF to the SML model [56] is now considered. Let the extended state vector be defined as

$$\xi = [S, X, K_b, K_c, K_m] \tag{72}$$

It consists of the two process variables (substrate and biomass) plus the three kinetic parameters K_b, K_c, K_m. If the system parameters are assumed to be constant in time, their dynamics can be expressed as $d\xi_i/dt = 0$ with $i = 3, 4, 5$. Using the new notation (72) in Eqs. (4), (5) yields

$$d\xi_1/dt = -\xi_1\xi_2\xi_3 - q(1 + r)\,\xi_1 + qS_i \tag{73}$$

$$d\xi_2/dt = \xi_4 - q(1 + r)\,\xi_2 - \xi_5\xi_2^2/\xi_1 + rqX_r \tag{74}$$

$$d\xi_3/dt = d\xi_4/dt = d\xi_5/dt = 0 \tag{75}$$

Considering that the updating measurement is available at discrete intervals h, two kinds of recursive equations are needed to mechanize the continuous — discrete filter:

1) Extrapolation of system Eqs. (73)–(75) between measurements at time t — h and t. During this interval no external information is available and both the system variables and the associated covariance matrix need to be extrapolated. The quantities $\hat{\xi}_i(\tau \mid t - h)$ denote the i-th estimated variable at time $t - h \leqq \tau < t$ based on the measurements up to time t — h, whereas the system dynamics **f**(.) is linearized for computing the covariance matrix i.e. $\mathbf{A} = \partial \mathbf{f}/\partial \xi$.
Filter extrapolation:

$$d\hat{\xi}_i(\tau \mid t - h)/dt = \mathbf{f}(\hat{\xi}(\tau \mid t - h), S_i(\tau)] \; ; \tag{76}$$

covariance matrix extrapolation:

$$d\mathbf{G}(\tau \mid t - h)/dt = \mathbf{AG}(\tau \mid t - h) + \mathbf{G}(\tau \mid t - h)\,\mathbf{A}^T + \mathbf{W} \tag{77}$$

2) Updating of filter quantities (state variables and covariance matrix) are performed as soon as the new measurement becomes available at time t.

Filter update:

$$\mathbf{K}_t = \mathbf{G}(t \mid t - h)\,\mathbf{c}^T(\mathbf{cG}(t \mid t - h)\,\mathbf{c}^T + \mathbf{R})^{-1} \tag{78}$$

$$\hat{\xi}_i(t \mid t) = \hat{\xi}_i(t \mid t - h) + \mathbf{K}_T(m_t - \mathbf{c}\hat{\xi}_i(t \mid t - h)) \; ; \tag{79}$$

covariance matrix update:

$$\mathbf{G}(t \mid t) = \mathbf{G}(t \mid t - h) - \mathbf{K}_T\mathbf{cG}(t \mid t - h) \tag{80}$$

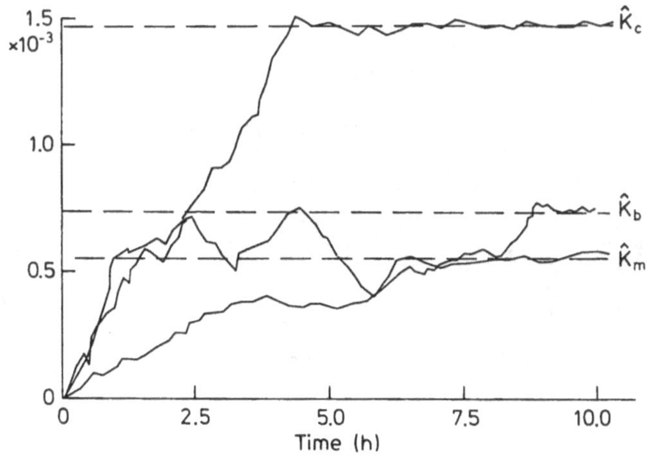

Fig. 12. Extended Kalman filtering of the SML model. Substrate and biomass data from the municipal medium scale plant were used to estimate the three parameters $\{K_b, K_c, K_m\}$. The dashed lines indicate the parameter values obtained with the off-line algorithm Eqs. (60)–(67)

where m_t is the t-th measurement and $\mathbf{K}_t$ is the corresponding Kalman gain used to update the filter. Likewise, $\mathbf{G}(t \mid t - h)$ and $\mathbf{G}(t \mid t)$ represent the covariance matrix before and after the updating of the t-th measurement. Lastly, $\mathbf{W}$ and $\mathbf{R}$ are covariance matrices depending on the statistical characteristics of the data, and $\mathbf{c}$ is a vector relating the filter state variables to the output (measured) variable. The algorithm (73)–(80) was applied to the SML model to estimate the three system parameters $\{K_b, K_c, K_m\}$ in addition to the two system variables $\{S, X\}$ using the same experimental data set from the municipal medium scale plant from which the estimates of Table 2 were obtained. The estimation results are shown in Fig. 12. It can be seen that the EKF algorithm converges to the same values which were obtained with the off-line algorithm Eqs. (60)–(67) with the difference that in this arrangement the measurements are processed one at a time, hence this algorithm can be used for on-line process monitoring. A further EKF application to the SML model is reported in [56] where dissolved oxygen experimental data are used to estimate four system parameters $\{K_b, K_c, K_m, K_La\}$ together with the three process variables $\{S, X, C\}$.

In this limited information experiment the EKF approach is still feasible, as discussed in [56], although its estimation performance is slightly degraded. Fig. 13 shows the estimation results for this case, in which an experimental set of 48 hourly DO data was used. Though the convergence is slower, a suitable choice of the filter parameters $\mathbf{W}$ and $\mathbf{R}$ produces stable and reliable estimates.

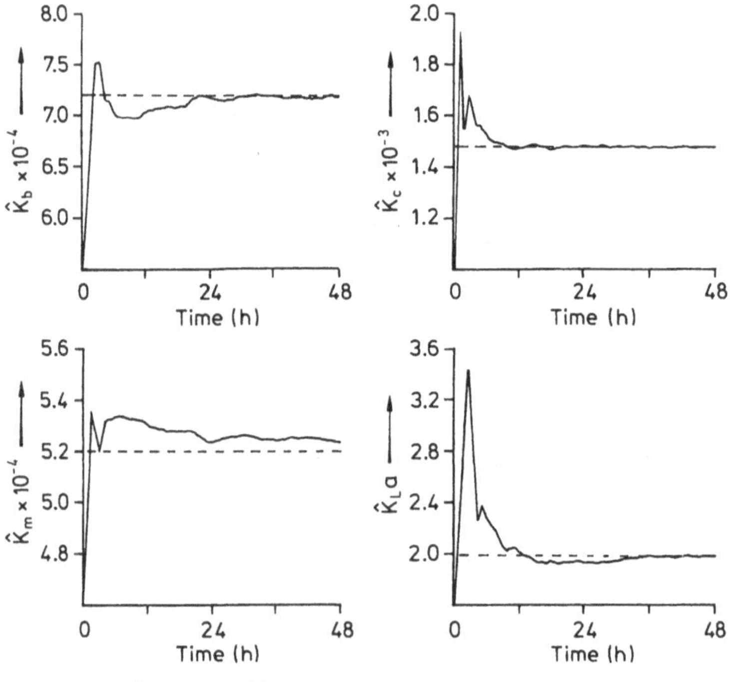

Fig. 13. Extended Kalman filtering of 48 hourly dissolved oxygen data to estimate the four system parameters $\{K_b, K_c, K_m, K_La\}$ [56]

3.5 On-line Estimation of Bioactivities

It has already been shown that oxygen uptake rate (OUR) is a key process quantity, from which valuable information on the entire process can be inferred. Then it is not surprising that the estimation of this variable has received considerable attention. The importance of OUR as an indicator of bioactivities was first assessed by Olsson and Andrews [18] and Stenstrom and Andrews [19] and several results became available soon after regarding the estimation of OUR and its companion parameter K_La, indicating the ability of the mixed liquor to accept oxygen from the adjacent gaseous phase. Given the dissolved oxygen dynamic balance

$$dC/dt = K_La(C_{sat} - C) - OUR \tag{9}$$

solving for OUR would appear a straightforward task, if it were not for the uncertainty on K_La (either unknown or time-varying) and the numerical difficulty in obtaining a reliable approximation of dC/dt. If either or both problems can be tackled independently, then partial solutions to Eq. (9) can be found, as described by Holmberg [39, 42], Marsili-Libelli [57], Goto and Andrews [58], Howell and Sodipo [59]. The common endeavour is to recast the non linear estimation problem in the context of linear filtering theory and apply meaningful results, such as the extended Kalman filter [53, 54]. Holmberg [39] addressed the problem thoroughly and proposed two alternative approaches. If K_La is known through prior experiments and DO is controlled to a given set-point, then Eq. (9) yields a straightforward solution. Conversely, if the DO level is allowed to vary and if K_La is expressed as a function of the air flowrate, then a recursive least-squares estimator can be set-up to estimate both OUR and K_La. However, expressing K_La as a function of the air flow produces biased estimates, unless very special test conditions are met. Another recent result [58] is still based on a preliminary off-line determination of K_La from which OUR is obtained through Eq. (9) either with suitable numerical approximation of dC/dt or by time discretization. Holmberg and Olsson [60] describe a simultaneous estimation scheme for K_la and OUR based on a linear Kalman filter and taking advantage of the differing time scale of the two variables: in fact the oxygen utilization may vary significantly in a matter of hours, whereas K_La variations can be detected over several days. Moreover, care is placed in approximating the derivative dC/dt. The DO dynamics can be put in a simpler form, choosing the dissolved oxygen deficit (DOD) C(t) as a new variable

$$dD(t)/dt = -K_LaD(t) + OUR(t) + q(1 + r)\,C(t) \tag{81}$$

sampling Eq. (81) at h intervals, yields the discrete-time equivalent

$$D(t + h) = aD(t) + bOUR(t) + bq(1 + r)\,C(t) \tag{82}$$

where

$$a = \exp(-hK_La) \quad \text{and} \quad b = \frac{1}{K_La}(1 - a) \tag{83}$$

provided that both OUR and C do not vary appreciably during the sampling interval h. If K_La is known a priori, then Eq. (82) can be solved to yield OUR

$$OUR(t) = \frac{1}{b}\{D(t + h) - aD(t)\} - q(1 + r)\,C(t) \tag{84}$$

Otherwise, if both OUR and K_La have to be estimated concurrently, the bilinear nature of Eq. (82) prevents a successful application of the classical recursive least squares. In fact, parameter b is a function of K_La and the term b OUR is the product of the two estimated quantities. This difficulty, which has not been sufficiently highlighted in the literature, can be circumvented by expanding a from Eq. (83) in power series, retaining only the linear term and substituting in b to obtain $b \simeq h$. In this way decoupling is achieved and recursive least squares can be applied to the approximate system

$$D(t + h) = aD(t) + hOUR(t) + hq(1 + r)\,C(t) \tag{85}$$

From Eq. (85) the estimation error equation can be determined as

$$z(t) = \mathbf{p}(t)^T\,\mathbf{f}(t) + e(t) \tag{86}$$

where $e(t) = [\mathbf{p} - \hat{\mathbf{p}}(t)]^T\,\mathbf{f}(t)$ is the error induced by parameter mismatch and the following quantities are defined.

$$\text{Parameters:} \qquad \mathbf{p}(t)^T = [aOUR(t)]\;; \tag{87}$$

$$\text{representers:} \qquad \mathbf{f}(t) = [D(t)\;h]^T\;; \tag{88}$$

$$\text{predicted output:} \quad z(t) = D(t + h) - qh(1 + r)\,C(t) \tag{89}$$

with the symbol (T) denoting transpose. The estimated parameter vector $\hat{\mathbf{p}}(t)$ can be updated according to the following recursive least squares scheme

$$\hat{\mathbf{p}}(t) = \hat{\mathbf{p}}(t - h) + \mathbf{G}(t)\,\mathbf{f}(t)\,e(t) \tag{90}$$

$$\mathbf{G}(t + h) = \frac{1}{L_1(t)}\left[\mathbf{G}(t) - \frac{\mathbf{G}(t)\,\mathbf{f}^T(t)\,\mathbf{G}(t)}{L_1(t)/L_2(t) + \mathbf{f}^T(t)\,\mathbf{G}(t)\,\mathbf{f}(t)}\right] \tag{91}$$

The two parameters $L_1(t)$ and $L_2(t)$ appearing in the covariance matrix update Eq. (91) are such that $L_1(t)/L_2(t)$ = constant and $L_1(t)$ is adjusted in such a way to make *trace* $[\mathbf{G}(t)]$ = constant. In this way, suggested by Landau and Lozano [61)], $\mathbf{G}(t)$ never vanishes thus ensuring that time-varying parameters will be tracked. This is a crucial requirement because both parameters, but especially OUR, are time-varying quantities. Several applications of the above scheme appeared in the literature already quoted. However, the main pitfall of this approach is that the tracking capability of the basic least squares estimator is inadequate to follow the short-term variations of OUR with sufficient accuracy. Including a constant term (h) in the representer vector

$\mathbf{f}(t)$ and pretending to estimate OUR as the corresponding time-varying parameter is a very crude approach, resulting in biased estimates. Even worse, the error induced by the wrong representer choice reflects on the constant parameter also, hence K_La too is incorrectly estimated. A partial solution to this problem, as suggested by Clarke and Gawthrop [62], is to select a sampling period h much shorter than the fundamental period of OUR. Further, it is dOUR/dt which ultimately determines the tracking limit, and therefore the amplitude too has to be taken into account.

A more robust approach for joint K_La and OUR estimation, based on expressing OUR as a power series of time, can be developed along the guidelines suggested by Xianya and Evans for tackling inaccessible disturbances [63]. Let OUR be represented by a time-series approximation

$$\mathrm{OUR} = b_0 + b_1 t \tag{92}$$

where the coefficients $\{b_0, b_1\}$ are constant. Then the estimation scheme Eqs. (90), (91) can be used with the following parameter and representer vectors

$$\mathbf{p}^T = [a\ b_0\ b_1] \tag{93}$$

$$\mathbf{f}(t) = [D(t)\ 1 t_1]^T \tag{94}$$

where the new time variable t_1 is defined within one sampling interval h, as

$$t_1 = t - (k-1)h \quad \text{for} \quad (k-1)h \leqq t < kh \tag{95}$$

where k is an iteration counter. To avoid large numbers in the representers vector and to adjust the linear approximation as time progresses, the estimates are updated as follows

$$\hat{b}_0(kh) = \hat{b}_0(kh - t_1) + h\hat{b}_1(kh - t_1) \tag{96}$$

$$\hat{b}_1(kh) = \hat{b}_1(kh - t_1) \tag{97}$$

Between two adjacent resets at times $(k-1)h$ and kh the inaccessible quantity can be expressed by the linear approximation

$$\mathrm{O\hat{U}R}(t) = \hat{b}_0[(k-1)h] + t_1\hat{b}_1[(k-1)h] \quad \text{for} \quad (k-1)h \leqq t < RT \tag{98}$$

It might be argued that a higher order approximation including higher powers of t would give a better approximation, but two considerations suggest to keep the order low. First, increasing the order would imply estimating more parameters, resulting in a greater computational burden. This would in turn require a longer resetting period and therefore large numbers, likely to cause numerical problems. Figure 14 shows the joint estimation of OUR and a time-varying K_La over a 24-h period using the SML model in connection with the above algorithm Eqs. (92)–(98) and Square-Root [64, 65] factorization to guarantee positive definiteness of the covariance matrix **G**. The basic sampling interval is 3.75 min and resetting occurred at every four samples, i.e. at 15 min intervals.

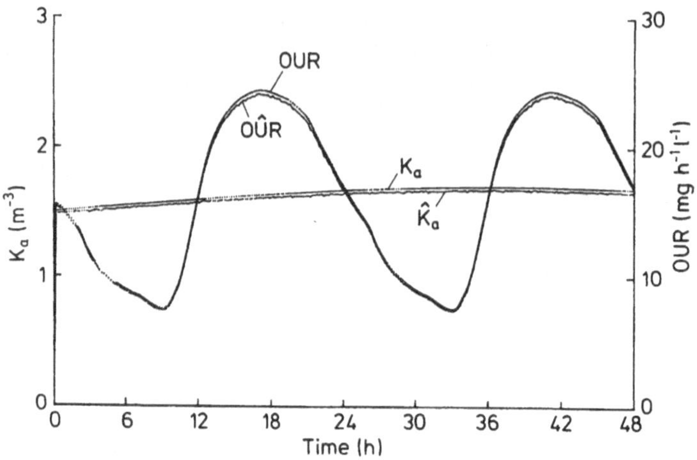

Fig. 14. Performance of the estimator Eq. (98) using the linear piecewise approximation Eqs. (92)–(97) for the reconstruction of the inaccessible input OUR and of a time-varying oxygen transfer coefficient K_La

3.6 On-line Estimation of Process Variables

Having established the necessary link between dissolved oxygen measurements and OUR, the analysis can be taken even further with the design of a state estimator to make for the unavailability of direct process measurements. Joint state and parameter estimation has been pursued by a number of researchers using the augmented state equations to derive asymptotically stable observers. The EKF approach can in fact be viewed in this light. Conversely, Aborhey and Williamson [38] derived a parameter estimator for the Monod kinetics under the assumption that both substrate and biomass measurements were available. Though of system-theoretic value, these results are of course not directly applicable, as the availability of either or both state variables is at present unrealistic. By contrast, Holmberg and Ranta [42] proposed an indirect state estimator based on the oxygen uptake rate measurement and Monod kinetics, expressing OUR as a linear combination of synthesis and maintenance metabolisms

$$\mathrm{OUR} = \alpha_s \mu(S)\, X + \alpha_e X \tag{99}$$

where α_s and α_e are respiration coefficients similar to γ_s and γ_e in the OUR Eq. (8) and $\mu(S)\, X$ is the microbial growth function. If the microorganism kinetics is given by the linear combination of growth and death processes

$$dX/dt = (\mu(S) - K_d)\, X \tag{100}$$

then solving Eq. (99) for $\mu(S)\, X$ and substituting in Eq. (100) yields a biomass estimator based on OUR measurements

$$d\hat{X}/dt = (-\alpha_e/\alpha_s - K_d)\, \hat{X} + \mathrm{OUR}/\alpha_s \tag{101}$$

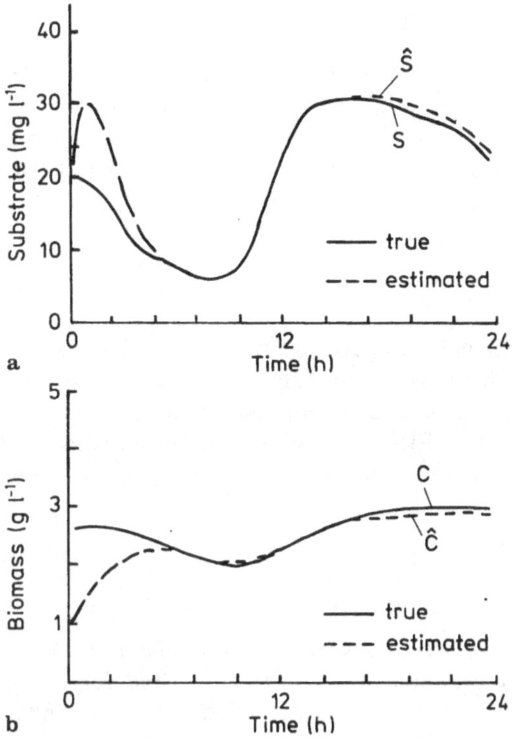

Fig. 15 a and b. Performance of the state estimator Eqs. (102), (103): **a** Estimated substrate $\hat{S}$, **b** estimated biomass $\hat{X}$

The above derivation was based on a linear decay term $-K_dX$, and resulted in a linear biomass estimator independent of the corresponding substrate dynamics. A similar reasoning based on the SML model (4), (5), (9) can be carried out using the pertinent OUR definition (8), repeated here for convenience

$$OUR = \gamma_s K_b SX + \gamma_e K_m X^2/S \tag{8}$$

Solving for the growth term K_bSX and substituting in the microbial dynamics Eq. (5) yields the substrate/biomass observer

$$d\hat{S}/dt = -K_b\hat{S}\hat{K} - q(1 + r)\,\hat{S} + qS_i \tag{102}$$

$$d\hat{X}/dt = -K_m(1 + \gamma_s/\gamma_e)\,\hat{X}^2/\hat{S} + OUR/\gamma_s - q(1 + r)\,\hat{X} + q_rX_r \tag{103}$$

Figure 15 shows the observer performance using DO data and assuming that OUR is directly available. If instead this variable is obtained from the algorithm just described by Eqs. (92)–(98) the accuracy is only slightly decreased. A sensitivity analysis of the observer Eqs. (102)–(103) is carried out in Refs. [57] and [86]. It was found that the biomass X is little affected by parameter inaccuracies, whereas the substrate S is more sensitive to variations of K_b, which on the other hand is the easiest to obtain experimentally or from literature data. Conversely K_m, the most difficult to tackle and the crucial one in the observer, has a much more limited influence.

As a concluding remark on this section, it can be stated that real-time parameter identification and reconstruction of process variables play a key role in activated sludge process control. In fact, given the inherent difficulty of obtaining direct process information, computer processing of simple measurements seems the only practical way to obtain an updated picture of the process development.

4 Process Control

The previous sections were concerned with modelling and identification of the activated sludge treatment process. These topics, in addition to having a scientific value in their own right, pave the way toward advanced process control needed to upgrade the performance in terms of reliability, treatment efficiency, and energy conservation. To achieve these goals the activated sludge process control system would have to overcome the following obstacles:

a) large variations of input organic load;
b) variations of sludge inventory and hence of kinetics;
c) limited availability of on-line process measurements.

These three aspects have already been addressed to justify the development of process models of limited complexity and their use for real-time estimation and control. It is the purpose of this section to survey the trends in activated sludge process control and to use the previous results for the design of efficient real-time controllers.

4.1 Process Performance Indicators

Most activated sludge plants operate in a time-varying environment and their operating conditions may be significantly different from the original design specifications. This may reduce treatment efficiency unless the plant operation is continually adjusted. Furthermore, operating costs have to be kept as low as possible without impairing the average effluent quality. Automatic plant control can be used to improve the operational performance and increase flexibility at low cost, but given the complex plant dynamics and differing time horizons, care must be placed in avoiding control actions which seek to improve the efficiency in the short term at the expenses of long-term performance. This brings into focus the need to design control laws which guarantee not only a consistent effluent quality, but also safeguard plant operation in the long run. For example, such process failures as "sludge bulking" (i.e. poor sludge settling quality), anoxic conditions or secondary settler denitrification may be caused by excessive aeration, possibly put into action to bring about a temporary improvement. Before undertaking any control law design, some meaningful biological process indicators must be defined. The most widely used of such indicators were first introduced by Lawrence and McCarty [66]. They are the food-to-mass ratio F_m.

$$F_m = \frac{\text{substrate mass utilized in 24-h}}{\text{bacterial mass in the aerator}} = \frac{\int_0^{24} (S_i - S)\, dt}{VX} \tag{104}$$

and the mean cell retention time, or sludge age, θ_c

$$\theta_c = \frac{\text{bacterial mass in the aerator}}{\text{bacterial mass wasted daily}} = \frac{VX}{wQX_r} = \frac{\theta X}{wX_r} \tag{105}$$

Where $\theta = q^{-1}$ is the hydraulic retention time. These two quantities, being derived from steady-state assumptions, are normally used for activated sludge plant design, but are of little value during time-varying operation, when the importance of maintaining a prescribed growth rate has been demonstrated by Garrett [67] and a number of other investigators. The following relation exists between θ_c and growth rate, if the Monod kinetics is assumed,

$$\theta_c^{-1} = \mu(S) - K_d \tag{106}$$

However, this cannot be used for real-time control since θ_c is defined as an average quantity, normally on a 24 h basis. Bisogni and Lawrence [68] have shown experimentally that growth rate as measured by θ_c is functionally related to specific oxygen uptake rate (SCOUR), defined as the ratio of oxygen uptake rate (OUR) to the biomass concentration in the aerator. To obtain a relationship between SCOUR and growth rate it is not necessary to write a material balance around the process nor is it necessary to assume steady-state conditions. SCOUR relates directly to growth rate as follows [19]:

$$\text{SCOUR} = \frac{\text{OUR}}{X} = \frac{(1-Y)}{Y}\mu(S) + \gamma_e K_d \tag{107}$$

4.2 Conventional Control Strategies

The following section will revise the current developments in control design for the activated sludge process and adapt some well-established control theory results such as the proportional-integral-derivative (PID) controllers to this particular process. Early literature contributions have applied the direct optimization methods to the Monod kinetics obtaining a nonlinear optimal regulator [69-71]. Though a very elegant solution results, the complexity of the ensuing control law makes their implementation difficult.

4.2.1 PID Control

As far as practical regulators are concerned, the proportional-integral-derivative (PID) controller is the most widely used in wastewater engineering. Its properties and design techniques are well known [72-74]. The relationship between PID regulator input $e(t) = y_{sp}(t) - y(t)$ and its corresponding output u(t) is:

$$u(t) = K_p\left[e(t) + \frac{1}{T_i}\int_0^t e(\sigma)\,d\sigma + T_d\frac{de(t)}{dt}\right] \tag{108}$$

The three coefficients $\{K_p, T_i, T_d\}$ are usually referred to as proportional gain, reset time, and lead time respectively. The inverse of the reset time T_i is referred to as the reset rate and expresses the time after which the integral action duplicates the proportional action. The derivative term provides a rough forecasting of error trend and as such enables preemptive control action. The lead time T_d is the time interval by which the derivative action anticipates the effect of the proportional control.

When a plant is controlled by a digital computer, the control law Eq. (108) is usually implemented in incremental discrete-time, with sampling interval h, form to avoid numerical overflow. In this case the full values of the controlled variable y(t) and of the set-point $y_{sp}(t)$ are used and the integration is performed outside the incremental computation loop. Then the sampled PID is implemented in two steps:

Incremental control signal computation

$$\delta u(t) = K_p \frac{h}{T_i} y_{sp}(t) - K_p \left(1 + \frac{h}{T_i} + \frac{T_d}{h}\right) y(t) + K_p\left(1 + 2\frac{T_d}{h}\right) y(t-h) - K_p \frac{T_d}{h} y(t-2h); \tag{109}$$

full control signal computation

$$u(t) = u(t-h) + \delta u(t) \tag{110}$$

4.2.2 Approximate Optimal Control

The efficiency of the PID controller Eqs. (109), (110) depends on a sensible selection of the three coefficients $\{K_p, T_i, T_d\}$. Many design techniques exist for linear systems [72-74] which could be applied in this context if the process dynamics is linearized and possibly complemented with some performance criterion. Let the process kinetics be approximated by a set of linear discrete-time state and output equations:

$$\text{State equations:} \quad \mathbf{x}(t+h) = \mathbf{H}\mathbf{x}(t) + \mathbf{g}\mathbf{u}(t); \tag{111}$$

$$\text{output equation:} \quad \mathbf{y}(t) = \mathbf{c}\mathbf{x}(t) \tag{111 a}$$

where the substrate (S), biomass (X), and dissolved oxygen (C) are assumed as state variables, i.e. $\mathbf{x} = [S\ X\ C]^T$. The vector **c** depends on the selected output variable of the process and h is the sampling interval. The performance index to complement the linearized dynamics Eqs. (110), (111) may be defined as a weighted sum of the regulator inputs e(t) and outputs u(t):

$$\Gamma = \sum_{t=0}^{H_c} \{y^2(t) + \varrho u^2(t)\} = \sum_{t=0}^{H_c} \{\mathbf{x}^T(t)\, \mathbf{c}^T \mathbf{W} \mathbf{c} \mathbf{x}(t) + \varrho u^2(t)\} \tag{112}$$

where H_c is the control horizon, i.e. the number of control steps over which the control action is to be optimized, and $\mathbf{W} \geqq 0$ and $\varrho > 0$ are appropriate matrix and scalar weights to place the proper emphasis either on control accuracy (**W**) or control effort

(ϱ). The controller parameters are to be determined in order to minimize the performance index Eq. (112). Several options are available at this stage:

a) A linear feedback is selected $u(t) = \mathbf{F}\mathbf{x}(t)$ and the matrix $\mathbf{F}$ is determined in an optimal way from the Riccati equation associated with the Γ criterion and the system Eqs. (110), (111) [72].
b) A controller with known structure is selected and its parameters are determined by numerical means in order to minimize Γ, in much the same way as system parameters identification in Sect. 3.

It can be shown that between the two approaches there are deep similarities [75] and both have been applied to activated sludge control problems [76-80]. Given the linearized dynamics Eqs. (110), (111) and the quadratic performance index Eq. (112), to enhance the controller robustness integral action is incorporated into the system introducing the integral state η defined as

$$\eta(t + h) = \eta(t) + y(t) \tag{113}$$

The augmented system state vector is $\mathbf{Z} = [\mathbf{x}\eta]^T$ and its dynamics is obtained aggregating the state equations Eqs. (110) and (113) to yield the following matrices

$$\mathbf{A}^* = \begin{vmatrix} \mathbf{H} & 0 \\ \mathbf{c} & 1 \end{vmatrix}; \quad \mathbf{b}^* = \begin{vmatrix} \mathbf{g} \\ 0 \end{vmatrix}; \quad \mathbf{c}^* = [\mathbf{c} \quad 0] \tag{114}$$

The resulting Linear-Quadratic-Integral (LQI) [72, 75] control law is then of the form

$$u(t) = -\mathbf{F}_1\mathbf{x}(t) - f_2\eta(t) \tag{115}$$

If an infinite control horizon is assumed (i.e. $H_c \to \infty$), the matrix $\mathbf{F} = [\mathbf{F}_1, f_2]$ can be obtained through the backward recursion.

$$\mathbf{F}_j = (\varrho + \mathbf{b}^{*T}\mathbf{P}_{j+1}\mathbf{b}^*)^{-1}\, \mathbf{b}^{*T}\mathbf{P}_{j+1}\mathbf{A}^* \tag{116}$$

$$\mathbf{P}_j = \mathbf{A}^{*T}\mathbf{P}_{j+1}(\mathbf{A}^* - \mathbf{b}^*\mathbf{F}_j) + \mathbf{W} \tag{117}$$

which under very mild conditions on $\{\mathbf{A}^*, \mathbf{b}^*, \mathbf{c}^*\}$ converges to a non-negative-definite limit $\mathbf{P}^* \geqq 0$ and gain $\mathbf{F}^*$ to be used in Eq. (115) representing the optimal control law for the linearized system Eqs. (110), (111). The overall feedback structure is depicted in Fig. 16a where the external set-point y_{sp} is included. The complete closed-loop dynamics is then

$$\mathbf{Z}(t + h) = \mathbf{A}_c\mathbf{Z}(t) + \mathbf{b}_c y_{sp} \tag{118}$$

$$y(t) = \mathbf{c}^*\mathbf{Z}(t) \tag{119}$$

with system matrices $\mathbf{A}_c$ and $\mathbf{b}_c$ defined as follows

$$\mathbf{A}_c = \begin{vmatrix} \mathbf{H} - \mathbf{g}\mathbf{F}_1 & -\mathbf{g}f_2 \\ \mathbf{c} & 1 \end{vmatrix}; \quad \mathbf{b}_c = \begin{vmatrix} \mathbf{0} \\ -1 \end{vmatrix} \tag{120}$$

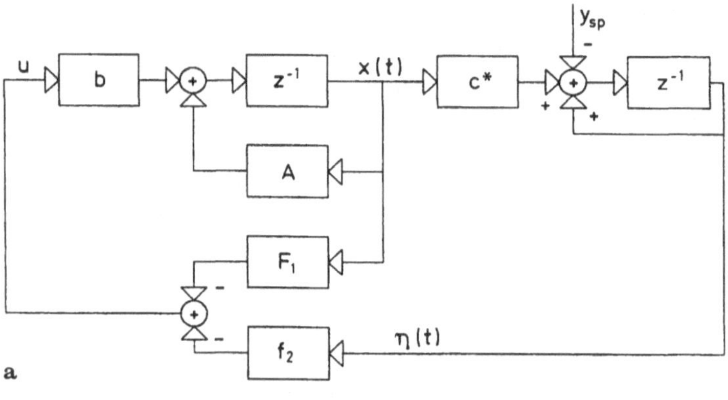

a

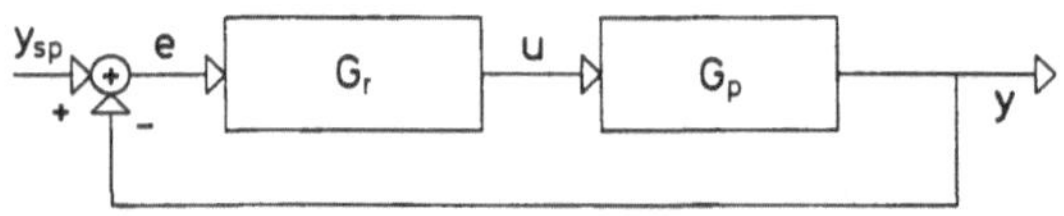

b

Fig. 16a and b. Structure of the discrete-time linear-quadratic-integral control scheme: **a** State feedback control incorporating integral action through matrix $\mathbf{F}_1$ and gain f_2, **b** equivalent output controller

It can be demonstrated [75] that the system Eqs. (118)–(120) can track a constant set-point even in presence of deterministic unmeasurable disturbances such as the time-varying organic load S_i. This controller, however, makes use of the entire augmented state $\mathbf{Z}(t)$, which is not a realistic assumption, but can be substituted by an equivalent output controller (Fig. 16b) using only dissolved oxygen measurements. The output controller can be determined from the closed-loop matrices $\{\mathbf{A}_c, \mathbf{b}_c, \mathbf{c}^*\}$ using the well-known equivalence between system matrices and closed-loop transfer function [72]

$$G_c(z) = \mathbf{c}^*(z\mathbf{I} - \mathbf{A}_c)^{-1}\,\mathbf{b}_c = \frac{G_p(z)\,G_r(z)}{1 + G_p(z)\,G_r(z)} = \frac{y}{y_{sp}} \tag{121}$$

where $G_r(z)$ and $G_p(z)$ are the regulator and process transfer functions respectively. From Eq. (121) the controller $G_r(z)$ in terms of an input-output difference equation can be obtained. For practical implementation it is more convenient to determine the transfer function in the backward shift z^{-1} operator:

$$G_r(z^{-1}) = \frac{G_c}{(1 - G_c)\,G_p} = \frac{u(z)}{e(z)} = \frac{a_1 z^{-1} + \dots + a_n z^{-n}}{1 + b_1 z^{-1} + \dots + b_n z^{-n}} \tag{122}$$

which yields the difference equation

$$u(t) = b_1 u(t-h) + \dots + b_n u(t-nh) + a_1 e(t-h) + \dots + a_n e(t-nh) \tag{123}$$

Equation (123) determines the controller output u(t) at time t based on past inputs u(t — h), u(t — 2h), ..., u(t — nh) and control errors e(t — h), e(t — 2h), ..., e(t — nh) up to n previous sampling intervals. This approach yields a satisfactory performance in terms of control accuracy and has the advantage of including the control effort in the cost functional Eq. (112). The resulting controller Eq. (123) is very simple to implement and requires little additional hardware with respect to conventional automation. It is also easy to adjust the coefficients in order to avoid numerical problems. As an example, a successful implementation of the output controller (123) on a very limited microprocessor using integer arithmetics is reported in [75].

4.3 Activated Sludge Control

Activated sludge wastewater treatment processes may vary greatly in design. The relevant literature reflects this variety and a number of different control problems have been considered, covering all aspects from dissolved oxygen and specific oxygen utilization rate (SCOUR) control [18, 19, 39, 77–79, 87] to more comprehensive strategies involving sludge recycle [78, 80–86] and secondary settler management [88–90].

4.3.1 Dissolved Oxygen Control

For obvious reasons, DO control is by far the most studied aspect of the activated sludge process automation. It has been shown that the pertinent dynamics is very simple, at least so far as it is not required to structure the OUR term. Furthermore, unless the DO level is abnormally low, this dynamics is independent of the basic substrate-biomass interaction. A precise DO control has its own merits, assuring a good sludge quality, making available the right amount of oxygen needed by a time-varying biochemical demand, and avoiding undue energy expenditure. It is also a necessary prerequisite for any long-term plant control policy. Of course, the control objective depends on the kind of plant: in a complete-mix plant only the DO level would be controlled, whereas in a longitudinal plug-flow reactor [91] the shape of the DO profile [18] along the tank should be controlled.

Biological oxidation of carbonaceous substrate occurs when the dissolved oxygen concentration is above a threshold limit of about 1 mg l^{-1}. Since the oxygen utilization depends on the incoming organic load, dissolved oxygen control is required to maintain a constant DO value during these fluctuations. The relevant model is the DO dynamics in the aerator already established by Eqs. (8), (9) in Sect. 1.

$$\mathrm{OUR} = \gamma_s K_b S X + \gamma_e K_m X^2/S \; ; \tag{8}$$

$$dC/dt = K_L a(C - C_{sat}) - \mathrm{OUR} - q(1 + r)\, C_i \tag{9}$$

The manipulated variable is the airflow U_a and this quantity appears in Eq. (9) through the oxygen transfer rate coefficient K_La. Practical experience has shown that K_La varies with the process temperature

$$K_L a(T) = K_L a(T_o)\, 1.024^{(T - T_o)} \tag{124}$$

and a linear relationship exists between air flow and K_La [39, 87]

$$K_La = K_o + K_aU_a \tag{125}$$

For example, Holmberg found for the Suomenoja plant [39] $K_o = -0.011$ and $K_a = 0.0018$ if U_a is expressed as $m^3\ h^{-1}$ and K_La as h^{-1}. It was also found that the transfer efficiency is always less than that obtained with the same aeration equipment in tap water. Substituting Eq. (125) in Eq. (9) and neglecting the natural re-aeration K_o yields

$$dC/dt = (C_{sat} - C)\,K_aU_a - OUR - q(1 + r)\,C_i \tag{126}$$

It should also be remembered that the dissolved oxygen saturation value C_{sat} depends on the temperature as follows

$$C_{sat} = 14.161 - 0.3943T + 0.007714T^2 - 0.0000646T^3 \tag{127}$$

Equation (127) is valid for tap water. For domestic wastes the saturation value of the sludge liquor should be multiplied by a factor of approximately 0.95. Moreover, if a submerged aeration system is considered, the diffusers depth must be taken into account as this affects the oxygen solubility. Therefore the saturation value is determined as follows

$$C_{sat}^{sub} = C_{sat}(P_b + H_a - v)/(760 - v) \tag{128}$$

where:
P_b = Barometric pressure (mm Hg);
H_a = Hydrostatic pressure at diffusers depth (mm Hg);
v = Liquor vapour pressure at temperature T (mm Hg)

Equations (8), (124)–(128) form the basic model for dissolved oxygen control in a completely mixed activated sludge plant. A PID controller can be used to determine the amount of air supplied to the system U_a depending on the difference between the required and actual DO levels. Hence, referring to Eqs. (108)–(110) the following quantities are defined: $e(t) = C_{sp} - C$ and $u(t) = U_a$. The main practical problem encountered in dissolved oxygen control is that the great majority of operating plants do not have any provision for varying the air flow rate continuously and the aeration equipment is operated on an on-off basis. However, as the advantages of adjustable air flow begin to emerge, more and more plants are being upgraded. Flexible aeration systems comprise movable weirs to control the submergence of surface turbines or variable DC motors driven blowers if bottom diffusers are used. However, the practical air flow regulation conceals several engineering problems as any change in the air flow affects the efficiency of the aeration system. In the bottom aeration case the blowers/diffusers combination is designed for maximum efficiency at a given air flow and would operate at a lower efficiency if a different flow is required. Particularly, at low flow the static pressure may stop the air flow completely even though the blowers are running. On the other hand, the regulating range of the blowers may be too narrow for the control requirements; therefore two or more partial control loops may be

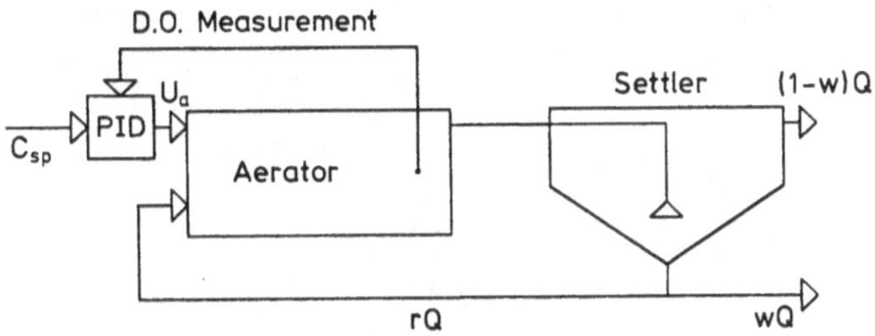

Fig. 17. Aeration control scheme using a PID regulator to manipulate the air flow and a dissolved oxygen probe in the aerator

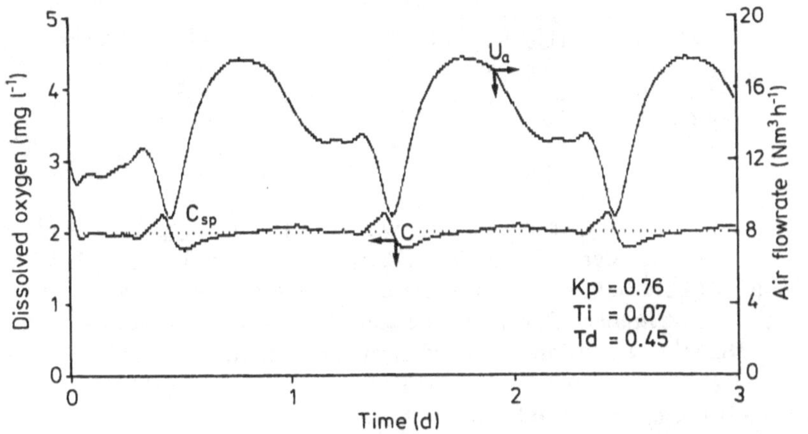

Fig. 18. Performance of the previous PID control scheme. The integral action is very strong to cope with the inaccessible OUR disturbance, whereas the derivative action is kept low to prevent high frequency oscillations in the controller

needed. Aarinen et al. [77] used a reciprocating compressor with continuously controllable DC motor drive. Personal experience has shown that since bottom diffuser require a substantial pressure to produce sufficiently fine bubbles, it is preferable to use a constant high pressure supply and to use a control valve to obtain the required flow. In the case of surface aeration, a limited degree of adjustment can be achieved by varying the submersion of the turbines and/or the speed of the drives, although usually the efficiency of the system depends critically on both factors, making the surface aeration systems less amenable to variable operation. The following examples assume that the plant is equipped with continuous air flow regulation.

Figure 17 shows a possible dissolved oxygen PID control scheme to maintain a set-point of 2 mg l^{-1}. Integral and derivative actions must be introduced to cope at least partially with the inaccessible disturbance represented by the variable incoming load. The simulation of Fig. 18 was obtained with the model of Sect. 2 and the incoming BOD varied from 90 to over 500 mg l^{-1} during the 24-h period. The PID coefficients were tuned by trial-and-error. It can be seen that the controller delivers a massive integral action ($T_i = 0.07$). This is required by the presence of the inaccessible dis-

Table 4. Operating conditions for optimal aeration control [79, 80]

Variable	Symbol	Operating value	Units
BOD	S	21	$mg\ l^{-1}$
Input load	S_i	350	$mg\ l^{-1}$
MLSS	X	2800	$mg\ l^{-1}$
Recycle MLSS	X_r	9600	$mg\ l^{-1}$
DO	C	2	$mg\ l^{-1}$
Input D.O.	C_i	1	$mg\ l^{-1}$
Airflow	U_a	12	$Nm^3\ h^{-1}$
Dilution rate	q	0.14	h^{-1}
Recycle rate	r	0.50	—

Parameter	Value	Units
K_b	7.42×10^{-4}	$l\ mg^{-1}\ h^{-1}$
K_c	1.47×10^{-3}	$l\ mg^{-1}\ h^{-1}$
K_m	5.23×10^{-4}	h^{-1}
K_a	0.2286	Nm^{-3}

turbance represented by the oxygen uptake rate which varies the oxygen demand of the system. By contrast, the derivative action is kept rather limited ($T_d = 0.45$) to avoid excessive high frequency oscillation in the control signal that would be detrimental to the aeration equipment. It can also be seen that the air flow rate has fairly large swings and that the regulation is imperfect at the beginning of the high load hours (from 10:00 a.m. to 1:00 p.m.) when the process conditions undergo a drastic change, which the PID can accommodate only in part.

By contrast, if the controller parameters were to be selected to satisfy some optimal criterion, the approach of Sect. 4.2.2 and the objective function Eq. (112) can be adopted. This approach has been applied to the pilot plant whose data were previously used to calibrate the SML model (see Table 2). The selected operating point is reported in Table 4 and the resulting continuous-time system matrices are

$$A^* = \begin{vmatrix} -2.3347 & -0.0155 & 0.0 \\ 13.9830 & -0.3212 & 0.0 \\ -0.4184 & -7.86 \times 10^{-3} & -2.954 \end{vmatrix} \qquad b^* = \begin{vmatrix} 0.0 \\ 0.0 \\ 1.52 \times 10^{-3} \end{vmatrix} \tag{129}$$

with eigenvalues $\lambda_1 = -0.43556$; $\lambda_2 = -2.22043$; $\lambda_3 = -2.954$, which guarantee a stable linear approximation. Assuming a sampling interval $h = 15$ min the following discrete-time output regulator is determined

$$\begin{aligned} U_a(t) = {} & 3.1U_a(t-h) + 3.53U_a(t-2h) + 1.72U_a(t-3h) \\ & - 0.26U_a(t-4h) + 0.044U_a(t-5h) + 0.01U_a(t-6h) \\ & + 749.8e(t-h) - 1871e(t-2h) + 1749.1e(t-3h) \\ & - 761.4e(t-4h) + 155.5e(t-5h) - 12.1e(t-6h) \end{aligned} \tag{130}$$

The application of the approximate optimal regulator Eq. (130) to maintain a dissolved oxygen concentration of 2 mg l^{-1} is shown in Fig. 19 where it is compared with a

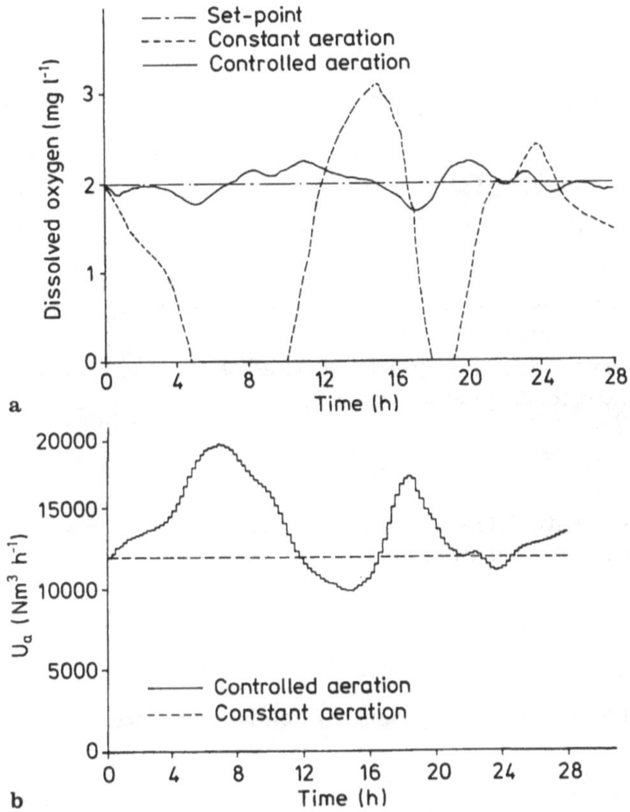

Fig. 19a and b. Optimal dissolved oxygen control using Eq. (130) (solid line) versus constant aeration (dashed line) applied to the pilot plant model for which the kinetic parameters were previously estimated (see Table 2) [80]

constant aeration system. It can be seen that the latter is not capable of supplying an adequate amount of oxygen during the peak loading hours. Apart from the avoidance of anoxic conditions, the energy saving obtained with the optimal controller are in the order of 30%. Similar energy savings have been reported in the literature for automatic control applications.

4.3.2 Sludge Recycle Control

Sludge recycle is the second most obvious control action and can be used to obtain certain biological conditions in the aerator [76,81] or for sludge conditioning in the secondary settler [88-90]. The control objective may have been determined through previous steady-state optimization, such as described by Lauria et al. [32], Keinath et al. [92], and Craig et al. [93]. Hence the control target may be a given food-to-mass ratio F_m or sludge age θ_c as defined by Eqs. (102), (103). To obtain this, Flanagan [82] proposed a steady-state recycle/waste control based on a Monod kinetics process model

1) *Aeration tank:*

$$\textit{substrate} \quad dS/dt = q(S_i - S) - \frac{\mu}{Y} X \, ; \tag{131}$$

$$\textit{biomass} \quad dX/dt = qrX_r - q(1 + r)X + (\mu - K_d)X \tag{132}$$

2) *Secondary settler:*

$$\textit{biomass balance} \quad Q(1 + r)X = Q(1 - w)X_e + Q(r + w)X_r \tag{133}$$

where X_e is the biomass concentration in the effluent. At steady state $dS/dt = dX/dt = 0$ and a mass balance at the secondary settler yields

$$(1 + r)X = (1 - w)X_e + (r + w)X_r \tag{134}$$

Combining Eq. (134) with the steady-state solution of Eq. (132) yields the waste flow ratio to maintain a prescribed solids retention time θ_c

$$w = \frac{XV\theta_c^{-1} - X_e}{X_r - X_e} \tag{135}$$

From the same Eqs. (131)–(134) the required recycle fraction is determined

$$r = X \frac{1 - q(\mu - K_d)}{X_r - X} \tag{136}$$

Equations (135) and (136) define a steady-state sludge management policy. Since the plant usually operates in a dynamic environment, it is necessary to complement these rules with automatic controllers if the time-varying nature of the operation is to be accounted for. The PID control scheme of Fig. 20 can be used, assuming that sludge density can be measured directly. Fig. 21 shows the performance of the PID obtained with trial-and-error controller tuning using the same daily-periodic influent load as in the simulation of Fig. 18 and assuming an abrupt set-point change from $X_{sp} = 2\,g\,l^{-1}$

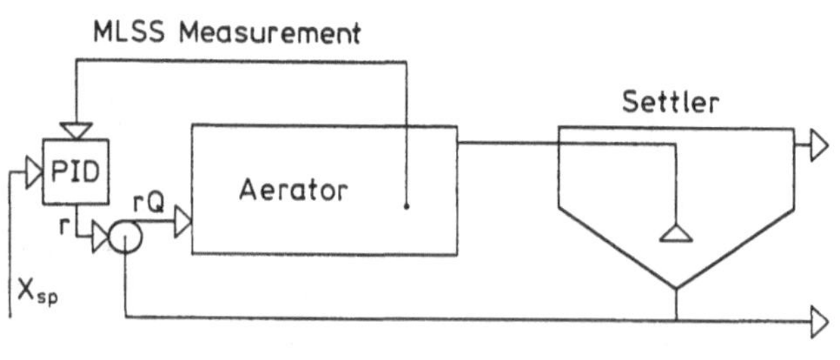

Fig. 20. Biomass control scheme using a PID regulator and assuming that biomass information is directly available

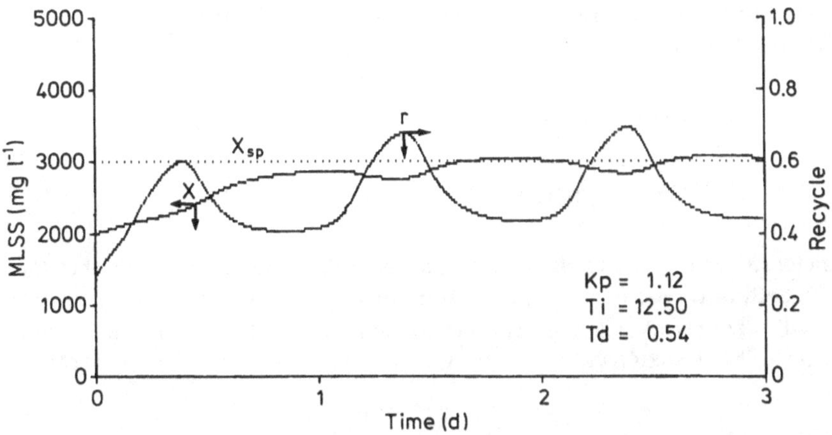

Fig. 21. Performance of the previous biomass control scheme. Contrary to the previous DO control, the integral action is small given the inherent "inertia" of the sludge dynamics. For the same reason the derivative action, acting as a forecast, has been increased

to $X_{sp} = 3\ g\ l^{-1}$. Contrary to the PID control of dissolved oxygen (Fig. 18) the reset time is very long ($T_i = 12.5$ h) indicating that the integral action is kept very low. In fact the sludge dynamics has such a large inherent "inertia" that a strong integral control would amplify its already sluggish response. Conversely, the short lead time ($T_d = 0.54$ h) indicates that derivative control is quite strong, as it must provide the predictive action required by the slow dynamics. In fact, it can be seen that the recycle diurnal pattern anticipates the daily fluctuation of the sludge concentration about the required set-point value.

As with dissolved oxygen control, the sludge control law can be complemented with an optimality index and its parameters determined accordingly. Several optimal control results are available in the literature: Hamalainen et al. [76] proposed a state feedback controller with a structure similar to Eq. (115) and a quadratic performance index as Eq. (112). In addition, the feedforward action was used to forecast the diurnal influent flow rate. Since the optimal state feedback law implies the availability of the entire state vector, as shown by Fig. 16a, a linear observer was used to obtain an estimate of BOD and biomass. The operational improvement of this solution was fully demonstrated.

A different approach was pursued by Sincic and Bailey [83], Yeung et al. [84], and Stehfest [85]. Their common assumption is that the plant input has a daily periodicity and the controller parameters are optimized to yield the best performance for that particular periodic input. This approach, which makes heavy use of simulation and numerical techniques, is well suited for plants dealing with domestic wastes, whose inputs are known to be consistently periodic. The resulting control law is strictly open-loop as it relies completely on the periodic assumption and cannot correct any departure from the assumed nominal pattern. Sincic and Bailey [83] consider the following performance index:

$$\Gamma = DT^{-1} \int_0^T Q(t)\,[1 + r(t)]\,S(t)\,dt + (1 - D)\,T^{-1} \int_0^T [S(t) - S_{av}]^2\,dt \quad (137)$$

where T is the optimization period usually assumed to be 24 h, and S_{av} is the average output BOD defined as

$$S_{av} = \frac{\int_0^T Q(t)[1 + r(t)]S(t)\,dt}{\int_0^T Q(t)[1 + r(t)]\,dt} \tag{138}$$

The parameter D can be used to shift the emphasis of the optimization on either term of Eq. (137). Setting D = 1 corresponds to minimizing the daily amount of substrate in the effluent whereas D = 0 corresponds to minimizing the fluctuations around S_{av}. Equation (138) was later used by Yeung et al. [84] to define another performance index

$$\Gamma = T^{-1} \int_0^T [S(t) - S_{av}]^2\,dt \tag{139}$$

The results so obtained indicate that this second approach yields better BOD removal that a feedback controller. The third periodic control exercise was recently proposed by Stehfest [85] assuming the daily average pollutant concentration as the performance index

$$\Gamma = T^{-1} \int_0^T (S + \alpha X)\,dt \tag{140}$$

where the parameter α is a weighting coefficient and the biomass X is included as an additional source of organic substrate. In [85] a comparison is made with a constant F_m recycle control to which the optimal periodic solution is clearly superior. As a concluding remark on periodic control, it must be emphasized that these results are not directly comparable because they were obtained with differing models, performance indexes,

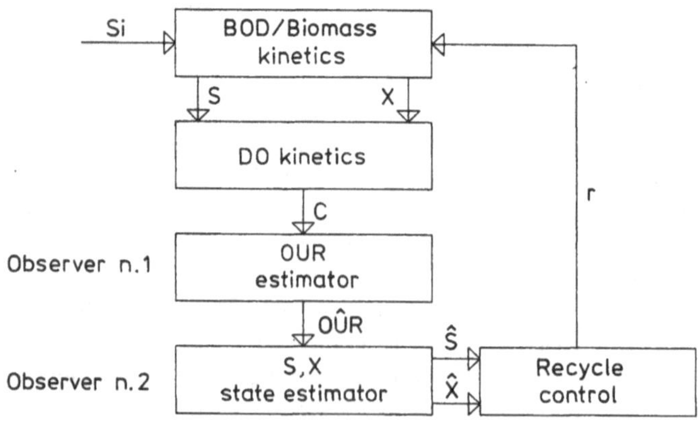

Fig. 22. Combined estimation and control scheme Eqs. (141)–(145) using dissolved oxygen measurements to implement set-point sludge control

and periodic input functions. The unavoidable weakness of the periodic approach is its feedforward philosophy: if for any reason the actual plant operation were to depart from the nominal pattern, the periodic controller would be incapable of any corrective action.

So far the problem of measurements availability has not been considered. In the case of periodic optimization this on-line information is immaterial, because the control strategy is determined off-line through iterative simulation and then implemented on the actual plant. Conversely, in feedback schemes it was so far assumed that biomass density could be directly measured by means of turbidimeters or ultrasonic devices. However, given the cost, complexity, and limited reliability of these on-line devices, it would be highly desirable to implement a sludge controller which makes use of inexpensive and reliable process measurements. For this, the observer theory developed in Sect. 3.6 can be used to obtain a controller based on dissolved oxygen measurements alone. This is made possible by the state observer (102)–(103) which relates the estimation of BOD and biomass to the oxygen uptake rate, which in turn can be reconstructed from dissolved oxygen measurements according to Eqs. (85) and (98). Mechanizing the double observer/controller scheme yields the aggregated structure shown in Fig. 22 where the first observer reconstructs OUR from DO measurements and supplies this information to the second observer which in turn estimates the unobservable state variables S and X, providing the regulator with their estimates $\hat{S}$ and $\hat{X}$. The complete double observer/controller scheme is then:

Observer n. 1: OUR estimation

$$\widehat{OUR}(t) = \hat{b}_0(t-h) + t_1\hat{b}_1(t-h) \quad \text{with} \quad 0 \leqq t_1 < h\,; \tag{141}$$

observer n. 2: substrate/biomass estimation

$$d\hat{S}/dt = -K_b\hat{S}\hat{X} - q(1+r)\,\hat{S} + qS_i\,, \tag{142}$$

$$d\hat{X}/dt = -K_m(1+\gamma_s/\gamma_e)\,\hat{X}^2/\hat{S} + \widehat{OUR}(t)/\gamma_s - q(1+r)\,\hat{X} + q_rX_r\,; \tag{143}$$

PID biomass controller:

$$\delta r(t) = K_p\frac{h}{T_i}X_{sp}(t) - K_p\left(1+\frac{h}{T_i}+\frac{T_d}{h}\right)X(t)$$

$$+ K_p\left(1+2\frac{T_d}{h}\right)X(t-h) - K_p\frac{T_d}{h}X(t-2h)\,, \tag{144}$$

$$r(t) = r(t-h) + \delta r(t) \tag{145}$$

Observer n. 2 is implemented in continuous-time for accuracy reasons, whereas the Observer n. 1 and the PID controller are discrete-time with sampling period h. The structure and performance of the combined scheme is analyzed in [86] and Fig. 23

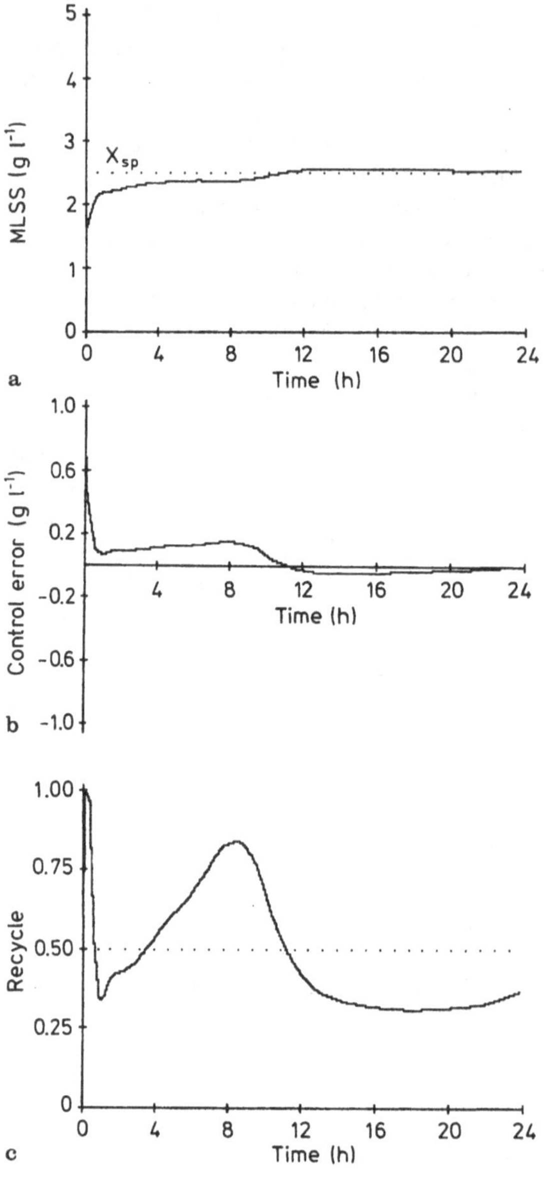

Fig. 23a–c. Performance of the combined estimation/control scheme of Fig. 22 in maintaining a required concentration X_{sp}: **a** Sludge concentration, **b** control error, **c** recycle fraction

shows its performance in the same conditions as previous schemes of Figs. 15, 17, and 20 with a biomass set-point $X_{sp} = 2.5$ g l^{-1}. The PID parameters are such that the desired set-point is reached after about 6 h and kept almost constant thereafter in spite of a variable influent organic load S_i. This satisfactory control action is achieved with an admissible control effort. In fact the recycle fraction r remains within the allowable limits (0, 1) (Fig. 23c). In practice a recycle greater than 1, though not representing a physical limit, should be avoided as it would imply a massive solids removal from the thickener underflow and a consequent sludge depletion. A sensitivity analysis of the combined scheme of Fig. 22 carried out in [86] revealed that the estimator/controller arrangement is only moderately sensitive to modelling errors, with the most sensitive parameter being K_La. Its sensitivity, however, is not so high to impair the overall efficiency.

4.3.3 SCOUR Control

The relationship between microbial activity and oxygen uptake rate has already been established through the specific oxygen uptake rate (SCOUR) defined as

$$SCOUR = \frac{OUR}{X} = \frac{(1 - Y)}{Y} \mu(S) + \gamma_e K_d \tag{107}$$

Since OUR is either directly measurable or can be calculated with the algorithm of Eq. (98) in Sect. 3.5, SCOUR is an on-line control variable whose validity has been demonstrated through computer simulation [19] and practical experience [87]. In fact, SCOUR is a meaningful growth indicator in a dynamically changing environment, whereas food-to-mass ratio F_m or mean cell retention time θ_c are not. Keeping SCOUR close to a given value by manipulating recycle flow and/or contacting pattern provides positive real-time control in presence of inflow disturbances. Each plant has an "optimal" SCOUR value depending on the kind of incoming sewage and microbial inventory. Once this value has been experimentally determined, a recycle control can be set up to obtain a sludge concentration X_{sp} compatible with the selected set-point $SCOUR_{sp}$

$$X_{sp}(t) = \frac{OUR(t)}{SCOUR_{sp}} \tag{146}$$

This value can be used as a set-point for a biomass PID control loop acting on the recycle flow rate as already considered in Fig. 20. The dynamic requirements of this loop are now more stringent since the PID is meant to track a variable $X_{sp}(t)$ given by Eq. (146), therefore the previous tuning cannot be carried over and new control parameters have to be selected.

Yust et al. [87] use SCOUR to compare recycle versus contacting pattern control policies. A step-feed and a completely mixed plants were run in parallel, a SCOUR value of 15 mg g^{-1} h^{-1} (O_2 per MLSS per time) was selected for both and the related sludge set-point $X_{sp}(t)$ was determined via Eq. (146). Then the control parameters were determined minimizing the following performance index

$$\Gamma = [X_{sp}(t) - X(t)]^2 \tag{147}$$

at each sampling interval, where X(t) is the sludge concentration in the most relevant process section. On-line measurements for sludge density in each compartment and oxygen uptake rate were provided. Figure 24 compares the real-time experimental results obtained manipulating the recycle flow in the completely mixed plant and redistributing the influent load in the step-feed plant. It can be seen that the latter exhibits a greater flexibility and follows the selected $SCOUR_{sp}$ set-point more closely.

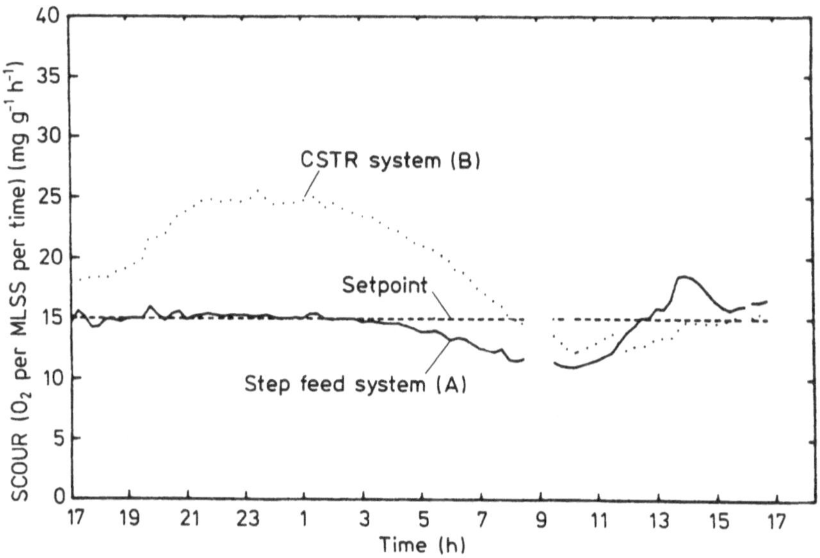

Fig. 24. SCOUR control of a step-feed pilot plant obtained by variable redistribution of the influent between two cascaded compartments. Reproduced by permission of The Institute of Measurement and Control on behalf of Yust et al. [87]

As a concluding remark, SCOUR control appears to be the most promising real-time control strategy as it provides the sludge controller with a target concentration related to the bioactivities in the process.

4.4 Adaptive Control

As already stated, the main difficulty in controlling biotechnological processes arises from the variability of kinetic parameters and the limited amount of on-line information. Though many significant applications of adaptive controllers are reported in the literature for specific biotechnological processes, very few results are as yet available for the activated sludge wastewater treatment process. The design and implementation of dissolved oxygen self-tuning controllers, the only available to date, will now be considered.

4.4.1 Structure of the Dissolved Oxygen Adaptive Controllers

Adaptive control is now a well-established branch of control theory (a thorough survey on adaptive controllers is presented in [94]). In this paper, the choice is restricted to a self-tuning regulator [62, 95]. Self-tuners can be divided in two broad categories: explicit and implicit, depending on whether the parameters of the process or of the controller are being estimated. For this application an implicit controller is considered, given the presence of the oxygen uptake rate, which is now regarded as an inaccessible deterministic disturbance. The literature on the applications of self-tuners to the activated sludge process is still very scarce. Apart from purely academic exercises, such as [96] in which OUR is unreasonably supposed constant, few other papers deal realistically with the practical implementation of dissolved oxygen control [97,98].

The self-tuning algorithm which is now introduced uses as little information as possible regarding process dynamics. The only practical assumption is that the aeration system is continuously adjustable. In other words, the plant is supposed to be equipped with some actuating mechanism to change the air flow rate between the allowable limits, such as movable weirs, variable submergence surface aerators or variable speed submerged blowers. The controller described in this section is a self-tuning version of the popular PID regulator, based on a modified algorithm proposed by Cameron and Seborg [99] and adapted to this context with nonlinear process dynamics and an inaccessible variable (OUR). The specific problem addressed in this section is set-point dissolved oxygen control, which was previously studied using conventional techniques. The limits of the conventional approach are clearly related to the inability of a predetermined controller to adjust to the changing operational environment. As in Sect. 4.2.1, the aim of the controller is to keep the DO level at a constant value C_{sp} in spite of input organic load fluctuations and/or parametric variations. The starting point for the controller design is again the sampled dissolved oxygen deficit (DOD) balance

$$D(t + h) = aD(t) + bOUR(t) + bq(1 + r)\, C(t) \tag{82}$$

where

$$a = \exp(-hK_La) \cong 1 - hK_aU_a \quad \text{and} \quad b = \frac{1}{K_La}(1 - a) \cong h \tag{148}$$

Decomposing Eq. (82) in the average and time-varying parts

$$U_a(t) = U_0 + \tilde{u}(t) \tag{149}$$

$$D(t) = D_0 + \tilde{d}(t) \tag{150}$$

$$C(t) = C_0 + \tilde{c}(t) \tag{151}$$

$$OUR(t) = OUR_0 + \widetilde{our}(t) \tag{152}$$

yields the following approximate linear static and dynamic balances

static balance:

$$K_aU_0D_0 = OUR_0 + q(1 + r)\, C_0\, ; \tag{153}$$

dynamic balance:

$$\tilde{d}(t + h) = (1 + a)\, \tilde{d}(t) - hK_aD_0\tilde{u}(t) + h\, \widetilde{our}(t) + hq(1 + r)\, \tilde{c}(t) \tag{154}$$

The linearized dynamic balance (154) is the starting point for the DO self-tuning controller design. It should be underlined that the $\widetilde{our}(t)$ term is supposed to be unknown,

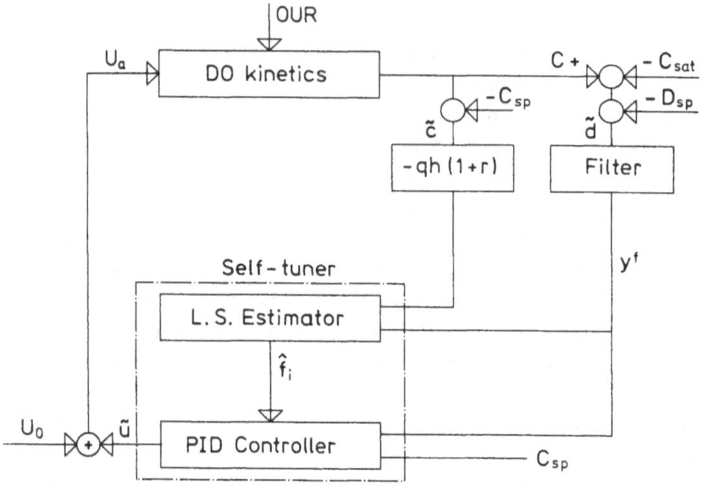

Fig. 25. Structure of the self-tuning PID regulator implementing dissolved oxygen control

therefore the controller should incorporate integral action to offset this inaccessible deterministic disturbance. Furthermore, the K_a parameter is generally unknown or time-varying, so there is a clear need for a self-tuning controller. The structure of the adaptive PID controller is depicted in Fig. 25, where its two main constituents (least squares estimator and PID regulator) are shown together with the inputs consisting of the filtered dissolved oxygen deficit (y^f) and the deterministic part $qh(1 + r)\,\tilde{c}(t)$.

4.4.2 Performance of the Self-tuning Controllers

The controller structure which produces the incremental air flow signal $\tilde{u}(t)$ has the following form

$$\tilde{u}(t) = \tilde{u}(t-h) + v\left[H^0 D_{sp} - \sum_{i=1}^{N} f_i y^f(t-ih)\right] \tag{155}$$

where:

D_{sp} is the dissolved oxygen deficit set-point: $D_{sp} = C_{sat} - C_{sp}$,

$\tilde{u}$ is the incremental manipulated variable (air flow rate),

y^f is the filtered DOD signal according to the recursive filtering equation,

$$y^f(t) = y^f(t-h)\,p_d + d(t) - D_{sp}\,, \tag{156}$$

f_i are the PID controller coefficients,

v is a weighting coefficient,

H^0 is a set-point weight defined so as to assure steady-state zero tracking error, i.e.

$$H^0 = \frac{1}{1 + p_d} \sum_{i=1}^{N} f_i \tag{157}$$

where $p_d \in (-1, 1)$ is the filter pole of Eq. (156),

N is the sum of the orders of the process model and of the filter Eq. (156).

The coefficients f_i are estimated on-line from past samples of the filtered output y^f using the following error equation

$$\varepsilon(t) = \tilde{d}(t) - \boldsymbol{\varrho}^T(t)\,\hat{\mathbf{p}}(t) - qh(1 + r)\,\tilde{c}(t) \tag{158}$$

where $\varepsilon(t)$ is the estimation error and the vectors of the representers $\boldsymbol{\varrho}(t)$ and of the estimated parameters $\hat{\mathbf{p}}(t)$ are defined as follows:

$$\boldsymbol{\varrho}(t) = [y^f(t)\; y^f(t-h) \ldots y^f(t-Nh)]^T \tag{159}$$

$$\hat{\mathbf{p}}(t) = [f_0, f_1, \ldots, f_N]^T \tag{160}$$

The estimation procedure is a recursive least-squares algorithm similar to Eqs. (90), (91) where the updating of the covariance matrix **G** is mechanized with the square-root [64,65] robust factorization method to avoid numerical degeneracy in the long run. More details on the derivation of this specific algorithm, including numerical tests and comparison with other self-tuning control laws can be found in [100]. These tests have shown that this regulator is fairly insensitive to the choice of the order N. Selecting $N = 2$, the result of Fig. 26 is obtained, where a variable and unknown K_a has been introduced. It can be seen that contrary to the deterministic algorithm of the previous Sect. 4.2, the self-tuner can adjust the air flow in order to keep the DO level as close as possible to the prescribed value DO_{sp} in spite of the K_a variations. It was also observed that increasing N beyond 2 did not produce any appreciable improvement. Conversely, the filter pole p_d and integral weight ν had a great influence on the performance. For the 48-h simulation of Fig. 26 the following values were selected:

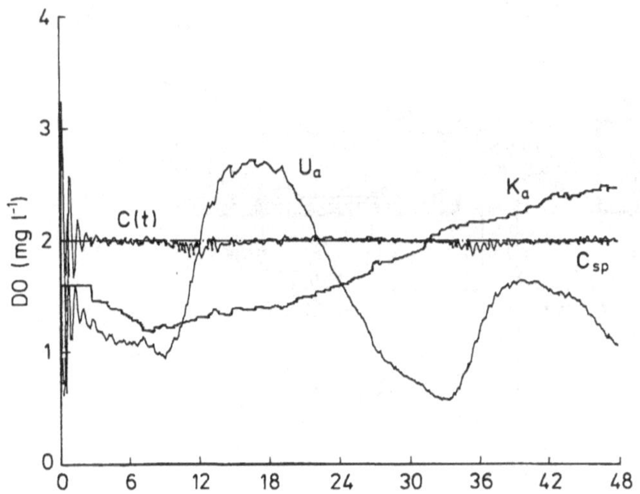

Fig. 26. Performance of the dissolved oxygen set-point self-tuning scheme of Fig. 25 with a variable and unknown K_La. The airflow adjusts to the changes of the transfer coefficient K_a

$p_d = 0.4$ and $v = 1.0$. The effect of a time-varying K_a on the controlled airflow U_a is quite evident: during the second 24-h period the controller decreases the aeration rate to adjust to an increased transfer efficiency K_a in order to maintain a given DO level C_{sp}.

As a concluding remark, the merits of this self-tuning PID scheme can be summarized as follows:

1) The controller is independent of any steady-state quantity such as the average air flowrate U_0. Therefore, no error in computing this preliminary parameter does impair the on-line operation.
2) The performance of the controller with a time-varying K_a is quite acceptable, at least with the rate of variation that is observed in practice.
3) The controller output is sufficiently smooth to avoid undue wear and tear of the actuating mechanism such as electrical drives and blowers.
4) The parameter estimates are very stable and do not degrade as time progresses, thus allowing unattended long-term operation.

It can be concluded that this PID self-tuning algorithm is suitable for the application to activated sludge plants equipped with adjustable aeration systems.

As anticipated, a similar approach was pursued by Olsson et al. [98] in producing the first engineering application of a dissolved oxygen self-tuning controller. The control scheme, implemented in a completely mixed activated sludge plant at Käppala (Sweden), relies on the usual dissolved oxygen model from which the linear sampled model is derived

$$C(t) = aC(t - h) + bU_a(t - h) \tag{161}$$

where the two unknown parameters {a, b} are estimated on-line with a least squares algorithm. The inaccessible deterministic disturbance OUR is not included in the

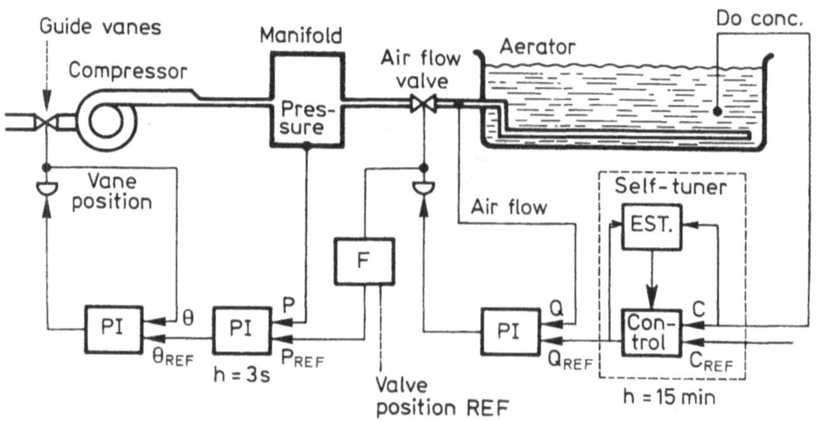

Fig. 27. Experimental self-tuning control scheme at Käppala plant showing the double control loop (guide vane and air flow valve) to adjust the airflow to the required value [98]. Published by the International Association on Water pollution Control (IAWPRC) in conjunction with Pergamon Press

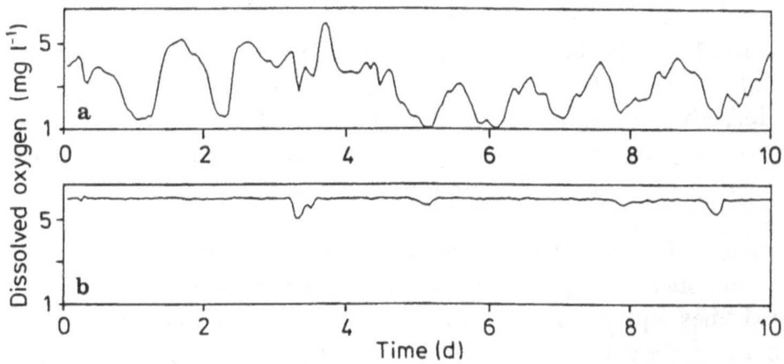

Fig. 28a and b. Experimental self-tuning control performance at Käppala plant [98]. Published by the International Association on Water pollution Control (IAWPRC) in conjunction with Pergamon Press. Dissolved oxygen concentrations with **a** manual adjustments, **b** self-tuning control

model and to compensate for this modelling error the controller has integral action. In fact the following sampled controller is selected

$$U_a(t) = U_a(t-h) - k_1[C(t) - C(t-h)] + k_2h[C_{sp} - C(t)] \qquad (162)$$

corresponding to a continuous-time proportional-integral (PI) action. Figure 27 depicts the Käppala controller implementation with three cascaded PI control loops to regulate the air flow. The self-tuner produces a reference air flow value Q_{ref} which is compared with the actual flow Q. The error is used to position the valve, which also depends on the pressure in the manifold and this second error is fed into the guide vanes controller. Using three cascaded PI controller enhances both flexibility and energy conservation, since the overall control action is aimed at minimizing the energy losses as the flow is varied. Figure 28 from [98] shows the performance of this arrangement in keeping the dissolved oxygen constant over a 7-day period. The authors also demonstrate that a constant DO level has improved the settleability properties of the sludge in the final clarifier.

5 Conclusions

This paper has surveyed the research advancements in mathematical modelling, estimation and control of the activated sludge wastewater treatment process. The presentation has focused on the structure and use of the SML reduced-order model to describe the interaction between carbonaceous BOD and biomass, with several other concurrent aspects of the process being included.

In Sect. 2, the simplified SML model has been introduced and its main structural features discussed. It was shown that it compares well with the widely used Monod kinetics, but has the advantage of being structurally simpler and numerically reliable. In addition to the main substrate-biomass interaction, other process streams were modelled, such as the oxidation of ammonium-nitrogen and the dynamics of the secondary clarifier.

Section 3 of the paper was devoted to the numerical calibration of the model, first dealing with its structural identifiability and then with the more practical aspect of making the best use of the available experimental information. It was concluded that the SML model is identifiable and can reproduce some well known experimental results with greater accuracy than the Monod model. On-line estimation of bioactivities was considered next, proposing an algorithm to estimate the oxygen uptake rate, a key process quantity.

The implementation of reliable control laws was considered in Sect. 4 and a review of several practical controllers was presented, including the popular PID regulator and a more advanced linear-quadratic-integral control. Some examples of practical controller design in connection with dissolved oxygen and sludge regulation were also presented. Lastly, self-tuning control has been considered as this approach is surely to draw more research work in the near future.

The aim of this paper was to demonstrate that modelling, estimation and control techniques can indeed improve the reliability and efficiency of the activated sludge process, whose capabilities have yet to be fully appreciated and exploited. The application of these control engineering tools can make up for the lack of sophisticated instrumentation and demonstrates that plant operation can be significantly improved by computer control with only a marginal increase over the total investment cost. The future trend is undoubtedly in this direction. A wider use of advanced control application is envisioned to obtain real-time process information, especially regarding biological indicators such as oxygen uptake rate and to implement reliable control strategies.

6 List of Symbols

A	secondary settler cross section [m^2]
C	dissolved oxygen concentration [$mg\ l^{-1}$]
C_i	input dissolved oxygen concentration [$mg\ l^{-1}$]
C_{sp}	dissolved oxygen set-point [$mg\ l^{-1}$]
C_{sat}	dissolved oxygen saturation value [$mg\ l^{-1}$]
D	dissolved oxygen deficit [$mg\ l^{-1}$]
F	total setting flux [$kg\ m^{-2}\ h^{-1}$]
F_m	food-to-mass ratio [$kg\ g^{-1}\ d^{-1}$] (BOD per MLSS per time)
h	sampling interval [h]
h_b	sludge blanket height [m]
K_a	air transfer coefficient [Nm^{-3}]
K_b	substrate decay rate (SML) [$mg^{-1}\ l\ h^{-1}$]
K_c	biomass specific growth rate (SML) [$mg^{-1}\ l\ h^{-1}$]
K_d	microbial decay rate (Monod) [h^{-1}]
K_m	biomass endogeneous metabolism rate (SML) [h^{-1}]
K_n	nitrifiers decay rate [h^{-1}]
K_o	dissolved oxygen half-velocity constant [$mg\ l^{-1}$]
K_s	half-velocity constant (Monod) [$mg\ l^{-1}$]
K_{am}	ammonium-nitrogen half-velocity constant [$mg\ l^{-1}$]
K_La	oxygen diffusion rate coefficient [h^{-1}]

M	accumulated biomass in the settler [kg]
N_a	nitrate-nitrogen concentration [mg l^{-1}]
n, a	batch flux settling curve parameters [—]
Q	process hydraulic flow rate [m^3 h^{-1}]
q	dilution rate [h^{-1}]
r, w	recycle and waste flow fractions [—]
S	substrate (BOD) concentration [mg L^{-1}]
S_i	input substrate (BOD) concentration [mg l^{-1}]
S_{am}	ammonium-nitrogen concentration [mg l^{-1}]
$S_{exp}, X_{exp}, C_{exp}$	substrate, biomass, and DO measurements [mg l^{-1}]
t, t_1	time [h]
U_a	air flow rate [Nm^3 h^{-1}]
u	settling sludge bulk settling velocity [m h^{-1}]
X	biomass concentration [mg l^{-1}]
X_a	settler sludge density above sludge blanket [mg l^{-1}]
X_b	settler sludge density below sludge blanket [mg l^{-1}]
X_e	sludge concentration in the effluent [mg l^{-1}]
X_n	nitrifying bacteria concentration [mg l^{-1}]
X_r	recycle biomass concentration [mg l^{-1}]
X_{sp}	sludge set-point [g l^{-1}]
Y	yield factor (Monod) [—]
Y_a	nitrifiers yield factor [—]
θ	hydraulic retention time [h]
θ_c	solids retention time (sludge age) [d]
$\hat{\mu}$	maximum specific growth rate (Monod) [h^{-1}]
$\hat{\mu}_n$	nitrifiers maximum specific growth rate [h^{-1}]
γ_e	oxygen utilization coefficient for endogenous metabolism [—]
γ_n	oxygen utilization coefficient for nitrification [—]
γ_s	oxygen utilization coefficient for synthetic activities [—]
ζ	feed height above settle bottom [m]

Acronyms:

BOD	biochemical oxygen demand [mg l^{-1}]
DO	dissolved oxygen [mg l^{-1}]
LS	least squares estimation algorithm
MLSS	mixed liquor suspended solids [mg l^{-1}], as a global measure of biomass in the activated sludge plant
OUR	oxygen uptake rate [mg l^{-1} h^{-1}]
SCOUR	specific oxygen uptake rate [mg g^{-1} h^{-1}]
SML	substrate/biomass reduced-order model, after the author's initials

Superscripts:

$\hat{}$	estimated variables
$\tilde{}$	incremental variable around a steady-state value
$()^T$	transpose of a matrix ()

7 References

1. Andrews, J. F.: Water Research *6*, 319 (1972)
2. Andrews, J. F.: ibid. *6*, 575 (1972)
3. Andrews, J. F.: ibid. *8*, 261 (1974)
4. Busby, J. B. and Andrews, J. F.: J. WPCF *47*, 1055 (1975)
5. Andrews, J. F.: Development of control strategies for waste-water treatment plants, in: Progress in Water Technology (eds. Andrews, J. F., Briggs, R. and Jenkins, S. H.), p. 233, Pergamon Press, Oxford 1974
6. Andrews, J. F., Stenstrom, M. K. and Burr, H. O.: Prog. Wat. Tech. *8*, 41 (1976)
7. Olsson, G.: Activated sludge dynamics I: Biological Models, Report 7511 (C), Lund Institute of Technology, p. 88 (1975)
8. Olsson, G.: AIChE Symposium Series, *72*, No. 159, 52 (1976)
9. Beck, M. B. (ed.): Operational water quality management: beyond planning and design, Executive Report 7, International Institute of Applied System Analysis (IIASA), Laxemburg (Austria), p. 74 (1981)
10. Marsili-Libelli, S.: Modelling and control of biological wastewater treatment, Trans. IMC *6*, No. 3, 115 (1984)
11. Marsili-Libelli, S.: Second Florence Workshop on modelling and control of biological wastewater treatment, Env. Tech. Letters, *6*, 515 (1985)
12. Roels, A. J.: Energetics and kinetics in biotechnology, p. 330, Amsterdam, Elsevier Biomedical Press 1983
13. Bazin, M. (ed.): Mathematics in microbiology, p. 307, London, Academic Press 1983
14. Marsili-Libelli, S.: Ecol. Mod. *9*, 15 (1980)
15. Marsili-Libelli, S.: ibid. *24*, 171 (1984)
16. Monod, J.: Ann. Rev. Microb. *3*, 371 (1942)
17. Poduska, R. A. and Andrews, J. F.: J. Water Pollut. Control Fed. *47*, 2599 (1975)
18. Olsson, G. and Andrews, J. F.: Water Research *12*, 985 (1978)
19. Stenstrom, M. K. and Andrews, J. F.: J. Env. Eng. Div. ASCE *105*, No. EE2, 245 (1979)
20. Vavilin, V. A.: Biotech. Bioeng. *24*, 8 (1982)
21. Vavilin, V. A. and Vasiliev, V. B.: ibid. *25*, 1521 (1983)
22. Vavilin, V. A. and Vasiliev, V. B.: ibid. *26*, 1042 (1984)
23. Vandevenne, L. and Eckenfelder, W. W.: Water Res. *14*, 561 (1980)
24. Esener, A. A., Veerman, T., Roels, J. A. and Kossen, N. W. F.: Biotech. Bioeng. *24*, 1749 (1982)
25. Henze, M. (ed): Modelling of Biological Wastewater Treatment, p. 204, Water Sci. Tech. *18* (1986)
26. Maynard-Smith, J.: Models in ecology, p. 146, Cambridge, Cambridge University Press 1974
27. Stenstrom, M. K. and Poduska, R. A.: Water Res. *14*, 643 (1980)
28. Gujer, W. and Erni, P.: Progr. Water Tech. *10*, 391 (1978)
29. Dick, R. I.: J. San. Eng. Div. ASCE *96*, No. SA2, 423 (1970)
30. Tracy, K. D. and Keinath, T. M.: AIChE Symposium Series (Water) *70*, No. 136, 291 (1973)
31. Shin, B. S. and Dick, R. I.: J. Env. Eng. Div. ASCE *106*, No. EE3, 505 (1980)
32. Lauria, D. T., Uunk, J. B. and Schaefer, J. K.: J. Env. Eng. Div. ASCE *103*, No. EE4, 625 (1977)
33. Stehfest, H.: Trans. IMC *6*, 160 (1984)
34. Olsson, G. and Chapman, D.: Modelling the Dynamics of Clarifier Behaviour in Activated Sludge Systems, in: Instrumentation and Control of Water and Wastewater Treatment and Transport Systems (ed. Drake, R. A. R.) p. 405, Oxford, Pergamon Press 1985
35. Heineken, F. G., Tsuchiya, H. M., and Aris, M.: Math. Biosci. *1*, 115 (1967)
36. Naito, M., Takamatsu, T., Fan, L. T., and Le, E. S.: Biotech. Bioeng. *11*, 731 (1969)
37. Nihtila, M. and Virkkunen, J.: ibid. *19*, 1831 (1977)
38. Aborhey, S. and Williamson, D.: Automatica *14*, 493 (1978)
39. Holmberg, A.: Int. J. System Sci. *12*, 703 (1981)
40. Homberg, A.: Math. Biosci. *62*, 23 (1982)
41. Solomon, B. O., Erickson, L. E., Hess, J. E., and Yang, S. S.: Biotech. Bioeng. *24*, 633 (1982)
42. Holmberg, A. and Ranta, J.: Automatica *18*, 181 (1982)
43. Di Stefano III, J. J. and Cobelli, C.: IEEE Trans. AC *25*, 830 (1980)

44. Pohjampalo, H.: Math. Biosci. *41*, 21 (1978)
45. Perkins, W. R.: Sensitivity Analysis, in: Feedback Systems (ed.) Cruz, J. B., p. 324, New York, McGraw-Hill (1972)
46. Vialas, C., Cheruy, A. and Gentil, S.: An Experimental Approach to Improve the Monod Model Identification, in: Modelling and Control of Biotechnological Processes (ed. Johnson, A.), p. 262, Oxford, Pergamon Press 1985
47. De Boor C. SIAM J. Numer. Anal. *14*, 441 (1977)
48. Himmelblau, D.: Applied Nonlinear Programming, p. 498, New York, McGraw-Hill 1972
49. Kuester, J. L. and Mize, J. H.: Optimization Techniques with FORTRAN, p. 500, New York, McGraw-Hill 1974
50. Marsili-Libelli, S. and Castelli, M.: Appl. Math. and Comp. *23*, n. 4, 341 (1987)
51. Bhatla, N. A. and Gaudy, A. F.: J. WPCF *38*, 1441 (1966)
52. Marsili-Libelli, S.: Envir. Tech. Lett. *7*, 341 (1986)
53. Jazwinski, A. H.: Stochastic Processes and Filtering Theory, p. 386, New York, Academic Press 1970
54. Gelb, D.: Applied Optimal Estimation, p. 374, Cambridge, Mass. M.I.T. Press 1974
55. Svrcek, N. Y., Elliot, R. F., and Zajic, J. E.: Biotech. Bioeng. *16*, 827 (1974)
56. Cook, S. and Marsili-Libelli, S.: Wat. Sci. Tech. *13*, 737 (1981)
57. Marsili-Libelli, S.: On-line Estimation of Bioactivities in Activated Sludge Processes, in: Modelling and Control of Biotechnical Processes (ed. Halme, A.), p. 316, Oxford, Pergamon Press 1983
58. Goto, M. and Andrews, J. F.: On-line Estimation of Oxygen Uptake Rate in the Activated Sludge Process, in: Instrumentation and Control of Water and Wastewater Treatment and Transport Systems (ed. Drake, R. A. R.), p. 867, Oxford, Pergamon Press 1985
59. Howell, J. A. and Sodipo, B. O.: On-line Respirometry and Estimation of Aeration Efficiencies in an Activated Sludge Aeration Basin from Dissolved Oxygen Measurements, in: Modelling and Control of Biotechnological Processes (ed. Johnson, A.) 191, Oxford, Pergamon Press 1985
60. Holmberg, U. and Olsson, G.: Simultaneous On-line Estimation of Oxygen Transfer Rate and Respiration Rate, ibid., 185
61. Landau, I. D. and Lozano, R.: Automatica *17*, 593 (1981)
62. Clarke, D. W. and Gawthrop, P. J.: Proc. IEE *126*, 633 (1979)
63. Xianya, X. and Evans, R. J.: ibid. *131*, Pt. D, 81 (1984)
64. Peterka, V.: Kibernetika *11*, 53 (1975)
65. Bierman, G. J.: Factorization Methods for Discrete Sequential Estimation, p. 241, New York, Academic Press 1977
66. Lawrence, E. W. and McCarty, P. L.: J. San. Eng. Div., ASCE *96*, n. SA3, 757 (1970)
67. Garrett, M. T.: Sew. and Ind. Wastes *30*, n. 3, 253 (1958)
68. Bisogni, J. J. and Lawrence, A. W.: Water Res. *5*, n. 9, 753 (1971)
69. D'Ans, G., Kokotovic, P. V., and Gottlieb, D.: IEEE Trans. AC *16*, 341 (1971)
70. D'Ans, G., Gottlieb, D., and Kokotovic, P. V.: Automatica *8*, 729 (1972)
71. Muzychenko, L. A., Macheva, L. A., Yakovleva, G. V.: Biotech. Bioeng. *4*, 629 (1974)
72. Kwaakernaak, H. and Sivan, R.: Linear Optimal Control Systems, p. 575, New York, Wiley-Interscience 1972
73. Stephanopoulos, G.: Chemical Process Control: an introduction to theory and practice, p. 696, Prentice Hall Int., Englewood Cliffs, N.J. 1984
74. Astrom, K. J. and Wittenmark, B.: Computer Controller Systems: theory and practice, p. 430, Prentice-Hall Int., Englewood Cliffs, N.J. 1984
75. Marsili-Libelli, S.: Int. J. of Control *33*, 601 (1981)
76. Hamalainen, R. P., Halme, A., and Gyllenberg, A.: A Control Model for Activated Sludge Treatment Process, 6th Triennial IFAC World Congress, ISA, Pittsburg 1975
77. Aarinen, R., Tirkkonen, J., and Halme, A.: Experiences on instrumentation and control of activated sludge plants — a microprocessor application, 7th Triennial IFAC World Congress, Pergamon Press, p. 255, 1978
78. Angelbeck, D. I., Shah Alam, A. B.: Simulation of optimal control strategies in a dynamic continuous flow activated sludge system, Proc. 30th Industrial Waste Conference Purdue University, Ann Arbor Science Publ. p. 159 (1977)

79. Marsili-Libelli, S.: Optimal aeration control for wastewater treatment, in: Computer Aided Design of Control Systems (ed. M. A. Cuenod), p. 511, Pergamon Press, Oxford 1980
80. Marsili-Libelli, S.: Trans. IMC *6*, No. 3, 146 (1984)
81. Sörensen, P. E.: Pilot-scale evaluation of control schemes for the activated sludge process, Water Quality Institute, Danish Academy of Technical Sciences, Horsholm, Technical Report N. 1, p. 249 (1979)
82. Flanagan, M. J.: AIChE Symposium Series *75*, No. 190, 232 (1979)
83. Sincic, D. and Bailey, J. E.: Water Res. *12*, 47 (1978)
84. Yeung, S. Y. S., Sincic, D. and Bailey, J. E.: Water Res. *14*, 77 (1980)
85. Stehfest, H.: Env. Tech. Letters *6*, 556 (1985)
86. Marsili-Libelli, S.: Water Sci. Tech. *16*, 613 (1984)
87. Yust, L. J., Stephenson, J. P., and Murphy, K. L.: Trans. IMC *6*, n. 3, 165 (1984)
88. Tsugara, H., Sekine, T., Fujimoto, E., and Matsui, S.: Prediction and control of resident sludge in a final clarifier, in: Instrumentation and Control of Water and Wastewater Treatment and Transport Systems (ed. Drake, R. A. R.), p. 653, Oxford, Pergamon Press 1985
89. Schlegel, S.: Control of the sludge gravity thickening process, ibid. p. 391, Oxford Pergamon Press 1985
90. Severin, B. F., Poduska, R. A., Fogler, S. P., and Abrahamsen, T. A.: Novel use of steady-state solids flux concepts for on-line clarifier control, ibid. p. 397, Oxford, Pergamon Press 1985
91. Levenspiel, O.: Chemical Reaction Engineering, p. 578, New York, John Wiley and Sons 1972
92. Keinath, T. M., Ryckman, M. D., Dana, C. H., and Hofer, D. A.: J. Env. Eng. Div. ASCE *103*, No. EE5, 829 (1977)
93. Craig, E. W., Meredith, D. D., Middleton, A. C.: J. Env. Eng. Div. ASCE *104*, No. EE6, 1101 (1978)
94. Harris, C. J. and Billings, S. A.: Self-tuning and adaptive control: theory and application, p. 333, IEE Press, London 1981
95. Astrom, K. J.: Automatica *19*, 471 (1983)
96. Yuh-ju Ko, K., McInnis, B. C., and Goodwin, G. C.: Automatica *18*, 727 (1982)
97. Cheruy, A., Panzarella, L., and Denat, J. P.: Multimodel simulation and adaptive stochastic control of an activated sludge process, in: 1st IFAC Workshop on Modelling and Control of Biotechnical Processes (ed. Halme, A.), p. 127, Pergamon Press, Oxford 1983
98. Olsson, G., Rundqwist, L., Eriksson, L., and Hall, L.: Self-tuning control of the dissolved oxygen concentration in activated sludge systems, in: Instrumentation and Control of Water and Wastewater Treatment and Transport Systems (ed. Drake, R. A. R.), p. 473, Oxford, Pergamon Press 1985
99. Cameron, F. and Seborg, D. E.: Int. J. of Control *38*, 401 (1983)
100. Marsili-Libelli, S., Giardi R., Lasagni, M.: Env. Tech. Letters *6*, 576 (1985)

FSC
www.fsc.org
MIX
Papier | Fördert gute Waldnutzung
FSC® C083411

Zeitfracht Medien GmbH
Ferdinand-Jühlke-Straße 7
99095 Erfurt, Deutschland
produktsicherheit@kolibri360.de